SpringerBriefs in Petroleum Geoscience & Engineering

Towards Sustainable Geoenergy and Transition to Net Zero

This series promotes and expedites the dissemination of substantive new research results, state-of-the-art subject reviews and tutorial overviews in the field of geoenergy exploration and production technology, extraction of geothermal energy, carbon capture, utilisation and storage (CCUS) and subsurface geoscience and engineering. The subject focus is on sustainable hydrocarbon production with an emphasis on the energy transition and achieving net zero emissions. These concise summaries (50-125 pages) will include cutting-edge research, analytical methods, advanced modelling techniques and practical applications. Coverage will extend to all theoretical and applied aspects of the field, including well drilling, shale-gas fracking, deepwater sedimentology, seismic exploration, pore-flow modelling and geoenergy economics.

Topics include but are not limited to:

- Sustainable hydrocarbon exploration and production
- Carbon capture, utilization and storage (CCUS)
- Geothermal energy
- Hydrogen production, transport, and storage
- Underground gas storage
- Application of artificial intelligence, machine learning and data analytics in geoenergy
- Exploration: Conventional and Unconventional
- Seismic Interpretation
- Formation Evaluation (well logging)
- Drilling and Completion
- Hydraulic Fracturing
- Geomechanics
- Reservoir Simulation and Modelling
- Flow in Porous Media: from nano- to field-scale
- Reservoir Engineering
- Production Engineering
- Well Engineering; Design, Decommissioning and Abandonment
- Flow Assurance, Mineral Scale & Hydrates
- Reservoir and Well Intervention
- Reservoir Stimulation
- Risk and Uncertainty
- Geoenergy Economics and Energy Policy

Contributions to the series can be made by submitting a proposal to the responsible Springer contact, Anthony Doyle: anthony.doyle@springer.com.

Chinmay Sethi · David A. Wood ·
Bodhisatwa Hazra · Mehdi Ostadhassan

Recent Advances in Shale Characterization Based on Laboratory Geochemistry and Geomechanics Techniques

 Springer

Chinmay Sethi
CSIR-Central Institute of Mining and Fuel
Research
Dhanbad, Jharkhand, India

Bodhisatwa Hazra
Coal Geology and Organic Petrology
Laboratory, Department of Applied
Geology
Indian Institute of Technology (Indian
School of Mines)
Dhanbad, Jharkhand, India

David A. Wood
DWA Energy Limited
Lincoln, Lincolnshire, UK

Mehdi Ostadhassan
State Key Laboratory of Continental
Shale Oil,
Northeast Petroleum University
Daqing, China

Institute of Geosciences, Marine and Land
Geomechanics and Geotechnics
Christian-Albrechts-Universität
Kiel, Germany

ISSN 2509-3126   ISSN 2509-3134 (electronic)
SpringerBriefs in Petroleum Geoscience & Engineering
ISBN 978-3-032-03960-6   ISBN 978-3-032-03961-3 (eBook)
https://doi.org/10.1007/978-3-032-03961-3

This Springer imprint is published by the registered company Springer Nature Switzerland AG
The registered company address is: Gewerbestrasse 11, 6330 Cham, Switzerland

If disposing of this product, please recycle the paper.

Contents

About the Author

Dr. Chinmay Sethi holds graduate and postgraduate degrees from Central University of Karnataka (India). He recently defended his Ph.D. thesis for his work on the organo-petrographic characteristics, source rock potential, and palaeo-depositional settings of shales of contrasting thermal maturities. Dr. Sethi's research interests include organic petrology, source rock geochemistry, geomechanics, and CO_2 storage. He has published multiple research articles in high-impact journals. His work has been presented at national and international conferences, where he has also received recognition, including a Best Paper Award at ICGEID 2024 hosted by IIT-Patna.

Dr. David A. Wood received a Ph.D. in geochemistry from Imperial College London, United Kingdom, in 1977. He held post-doctoral fellowships at Institut de Physique du Globe de Paris and University of Birmingham, focusing on ocean floor geochemistry and deep-water drilling. He has worked in energy industry operational, research, and consulting roles since 1980. In recent years, he has conducted focused research on natural gas, shales, carbonates, coals, kerogen kinetics, petrophysics, drilling, LNG, geodynamics, machine learning, carbon sequestration, hydrogen exploration and exploitation, renewable energy, and energy sustainability issues. He has published extensively (H-index >60) and served in editorial roles with several high-ranking scientific journals.

Dr. Bodhisatwa Hazra holds a Ph.D. from the Indian School of Mines, Dhanbad, India. He subsequently worked as a Postdoctoral Fellow at the Indian Institute of Technology, Bombay, followed by a tenure as DST-Inspire Faculty at the Indian Institute of Technology, Kharagpur. He earlier served as a Senior Scientist at the CSIR-CIMFR, Dhanbad, and is currently working as an Assistant Professor at the Coal Geology and Organic Petrology Laboratory, Department of Applied Geology, Indian Institute of Technology (Indian School of Mines), Dhanbad. Dr. Hazra's research expertise lies in organic geochemistry, organic petrology, petrophysics, CO_2

storage, and geomechanics. In 2019, he received the prestigious CSIR Young Scientist Award and, in 2023, the SERB International Research Experience (SIRE) Award to conduct research at the University of Kiel, Germany.

Dr. Mehdi Ostadhassan received his B.Sc. in petroleum exploration engineering from the Petroleum University of Technology (Iran), an M.Sc. in petroleum geophysics from IFP School (France), and a second M.Sc. in petroleum exploration engineering from PUT. He earned his Ph.D. in petroleum engineering from the University of North Dakota (USA). He began as Research Scientist at the Energy and Environmental Research Center and later joined the Department of Petroleum Engineering at UND, where he became Associate Professor in 2019. He then joined Northeast Petroleum University (China) as Distinguished Professor and is affiliated with the Unconventional Oil and Gas Institute. In 2021, he was awarded the prestigious Alexander von Humboldt Experienced Research Fellowship. He is currently associated with both Kiel University, Germany, and the State Key Laboratory of Continental Shale Oil, Northeast Petroleum University, Daqing, China. He secured $3M+ USD in public/private funding and collaborates globally with universities and private companies. He studies how subsurface forces influence fluids and organic/inorganic matter.

Chapter 1
Organic Matter (OM) and Dispersed Organic Matter (DOM) in Shales

Abstract The organic matter (OM) in shales is of two types, primary and secondary. Kerogen constitutes the primary OM; these are insoluble in organic solvents and comprise most of the OM present. As shales thermally mature over time during burial, the kerogen become progressively altered/transformed into secondary OM, particularly solid bitumen and pyrobitumen. This secondary OM typically occupies minute interstices between mineral grains. It is also associated with and promotes the formation of pores, at nano- to macro-scales, within the primary OM, providing space for storing hydrocarbons as they are generated during thermal maturation. That generated pore space can also be used to potentially store natural gas and possibly carbon dioxide injected into shale formations. The dimensions (quantity, size, and shape) of OM pores vary with the kerogen types present and are also influenced by thermal maturation degree and the surrounding inorganic minerals in the shale matrix. Imaging and gas-adsorption techniques make it possible to explore the evolution of OM-hosted porosity as thermal maturity advances and its influence on permeability, gas flow and storage. The paleoenvironmental conditions (water depth, oxygen levels, and rate of sedimentation) during shale deposition influences OM preservation. A thorough understanding of all these factors helps in delineating an effective shale gas reservoir for both hydrocarbon production and CO_2 sequestration.

Keywords Organic matter (OM) · Dispersed organic matter (DOM) · Kerogen types · CO_2 adsorption · Depositional environment · Paleoproductivity · Brittleness index · Correlative microscopy

1.1 Introduction

Soluble organic matter (SOM) and kerogen are two types of organic matter (OM) found in sedimentary rocks. The SOM is soluble in organic solvents, whereas kerogen is insoluble. 80 to 90% of the total organic matter in sedimentary rocks is kerogen

© The Author(s), under exclusive license to Springer Nature Switzerland AG 2025
C. Sethi et al., *Recent Advances in Shale Characterization Based on Laboratory Geochemistry and Geomechanics Techniques*,
SpringerBriefs in Petroleum Geoscience & Engineering,
https://doi.org/10.1007/978-3-032-03961-3_1

[56]. SOM and kerogen are known as primary OM as they are derived from the biological input and preservation conditions during sediment accumulation. Organic geochemists use the term "kerogen", whereas it is called "macerals" by organic petrologists. There are four types of kerogen (Type I, Type II, Type III, and Type IV) and are classified based on their origin, chemical composition, and hydrocarbon generation potential [39]. To classify kerogen types, van Krevelen diagram were used but it was developed originally for coal and it involved three kerogen types [24, 39, 56]. Modified van Krevelen diagrams are commonly used for analysis of OM characteristics (Fig. 1.1). This diagram plots hydrogen index (HI) against oxygen index (OI) derived from Rock–Eval pyrolysis to classify kerogen types [39]. The modified version of the diagram is considered the most cost effective and time-efficient approach.

On the other hand, in organic petrology, macerals, the microscopic components of OM, are classified into three main groups: vitrinite, inertinite, and liptinite [15, 16]. Secondary OM, in contrast to primary OM, forms through changes in primary OM during burial and thermal maturation [18]. It includes both portion that can be extracted and those that remain non-extractable, and are classified as pyrobitumen and bitumen. From organic petrology perspective, secondary OM comprises pyrobitumen, solid bitumen (SB), and oil. The term "solid bitumen" does not perfectly match the geochemical definition of "bitumen." Geochemists define bitumen as the organic material that can be dissolved and extracted from rock samples using solvents

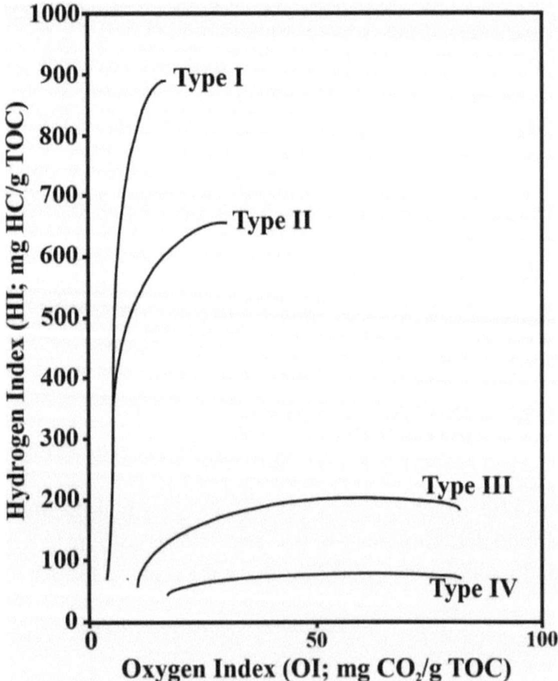

Fig. 1.1 Modified van Krevelen diagram depicting the relationship between HI and OI for determining the kerogen types. (Adapted from Geo-Marine Letters, Vol. 23, Langrock, U., Stein, R., Lipinski, M. and Brumsack, H.J., "Paleoenvironment and sea-level change in the early Cretaceous Barents Sea—implications from near-shore marine sapropels," pp. 34–42, Copyright (2003), with permission from Springer. License Number: 6040650037956)

like pentane or dichloromethane [60]. However, these terms partially overlap. In thermally mature rocks, such as shales in the peak oil window or beyond, the material extracted by organic solvents mostly originates from SB [8]. SB is most often found filling pores and fractures or forming accumulations aligned parallel to bedding planes. In whole rock samples, it can be identified microscopically based on its shape and mode of occurrence. SB, unlike primary macerals with their clear and defined structures, takes on the form of the surrounding space it resides in. However, when kerogen concentrates are prepared by acid digestion, the connection between the mineral matrix and the OM is lost, making it extremely challenging to differentiate SB in such samples. Another related term, "pyrobitumen," is referred by both petrologists and geochemists. It refers to the residue left behind after oil undergoes thermal cracking to form gas [53]. Nonetheless, pyrobitumen is not just the leftover from oil cracking, it is a type of SB which has undergone molecular cross-linking, making it insoluble [33]. As pyrobitumen is not soluble in solvents like carbon disulfide, it might be mistaken for kerogen if only the geochemical definition is used. However, under a reflected light microscope, pyrobitumen can be easily distinguished from kerogen. Pyrobitumen, like SB, fills the spaces between mineral grains, while kerogen macerals have distinct shapes of their own. These materials reflect the progressive thermal evolution of OM and its interaction with surrounding mineral matrices, contributing significantly to the reservoir quality of source rocks [33].

1.2 Role of DOM in Shales

Dispersed organic matter (DOM) controls the micro- and nano-scale behavior of shales such as hydrocarbon retention, migration and gas storage capacity [29]. Porosity of shales, an important parameter in determining reservoir quality, is affected by DOM. Thermal maturation leads to transformation of DOM by decomposition and generation of hydrocarbons which leads to development of pores in organic matrix known as secondary organic porosity [33]. The overall porosity of shales is enhanced due to this secondary organic porosity which ultimately improves the reservoirs gas storage capacity for hydrocarbons or CO_2. Studies carried out in recent years on the influence of DOM in CO_2 sequestration is gaining traction due to its impact on development of pores and adsorption capacity [1, 12, 37]. The factors affecting the adsorption and gas storage capacity of OM are its chemical composition, structure, and thermal maturity. High total organic carbon (TOC) and HI values correlate positively with abundant micropore and mesopore availability in shales [45]. CO_2 molecules demonstrate close affinity to OM due to its abundant microporous structure [25]. Thermal maturity plays a vital role in enhancing the CO_2 storage capacity of shale due to the associated pore development. Laboratory studies have observed that post-mature shales are more likely to exhibit higher adsorption capacity, as kerogen pores that were previously filled with bitumen are evacuated leading to development of additional space for gas storage [65, 70]. According to the International Union for

Pure and Applied Chemistry (IUPAC) organic pores are of three types: micropores (<2 nm), mesopores (2–50 nm), and macropores (50 nm) [55].

1.2.1   Organic Pore Morphologies and Characteristics: Insights from Imaging and Physisorption

Scanning electron microscopy (SEM) is being employed by numerous studies to investigate organic porosity in shales at micro- and nano-scales [29]. It is shown in different studies that size of organic pores is mostly 100 nm [41]. SEM cannot detect pores smaller than 2 nm. Also, it is difficult to relate the OM pores with specific OM type as OM appears dark under SEM because of its low atomic mass [29]. Optical-electron correlative microscopy is an emerging technique that uses optical microscopy to first detect the OM type, and then SEM is used to study its microstructural properties at higher resolution [9, 10, 50]. Hackley et al. [9] was first to employ this technique and demonstrated detailed characterization of OM which otherwise would not be possible by optical or electron microscopy alone.

Liu et al. [30] investigated OM porosity in New Albany shale of Ilionis basin with thermal maturity (Ro) ranging between 0.55 and 1.42% using similar technique. It was noted that vitrinite and inertinite macerals showed no evidence of porosity development with the increase of maturity. Some inertinite fragments were observed to show partially preserved cell lumens, which were filled with either authigenic quartz or SB. The study highlighted SB-hosted pores as the predominant form of OM porosity. Sethi et al. [50] through this technique studied the DOM in shales with contrasting thermal maturity and relate its microstructural characteristics as observed under SEM with the specific maceral type. Their method included adjustment of SEM parameters (acceleration voltage, working distance and brightness contrast levels) to clearly distinguish between the DOM and mineral phases. SEM-visible pores were observed to be present in sproinite found in immature shale, whereas sporinite in overmature shale did not show presence of pores. Both vitrinite and inertinite macerals in immature and overmature shales lacked SEM-visible pores.

SEM's inability to detect pores less than 2 nm in shales necessitates the use of complementary techniques such as low-pressure gas adsorption (using N_2 and CO_2 as adsorbates) for detailed understanding of pore parameters such as its volume, size, surface area, and distribution [7]. Research has shown that in shales, adsorption-derived micropore parameters and organic richness are positively correlated [10, 11, 45]. Marine shales are found to exhibit more extensively developed OM-hosted porosity compared to transitional or continental shales due to the prevalence of Type-I kerogen in marine shales [7]. On the other hand, transitional shales are observed to have less OM-hosted porosity compared to marine shales as they are generally composed of Type-III kerogen [6]. Lacustrine shales are composed of a mix of terrestrial and hydrogen-rich OM [21] and are characterized by OM pores that are less developed compared to marine shales [7].

Shale's pore structure is also influenced by the interaction between DOM and the surrounding mineral matrix. Bulk shale's overall porosity exhibits heterogeneity compared to that of isolated OM. Pores associated with minerals (clays, quartz, and carbonates) contribute to intricacies of shale pore network and impacts the rocks hydrocarbon flow and storage properties [7, 63]. Some rigid minerals such as quartz and carbonates act as framework minerals and prevent the pores found in OM and clay minerals (CM) from disintegration due to compaction [3]. Carbonate minerals are comparatively less effective in preservation of porosity compared to quartz [23]. Origin of quartz also impacts porosity evolution. Shale mineralogical composition dominated by biogenic silica results in a rigid siliceous matrix which contribute to effective pore preservation compared to terrigenous quartz [5, 35].

Additionally, CM are crucial for gas adsorption mainly in shale with >2% TOC [7]. Pores are found among the clay mineral particles and in between the mineral layers. Pore size, shape, specific surface area and surface roughness together affect the adsorption capacity of CM [37]. The intragranular pores, intergranular pores and microfractures are the predominant pore types in CM. Intergranular pores are primarily micro-sized which are formed by arrangement of clay particles in layers thereby referred to as stacked pores. These pores have wide distribution, a relatively large size, and good connectivity. Intragranular pores, on the other hand, are mostly interlayer nanoscale pores that are highly prevalent in CM. Their connectivity is generally moderate but can be high when linked to intergranular pores. Microfractures in shales generally characterized by micro-scale lengths and nanoscale widths and are formed due to diagenetic shrinkage and the clay and brittle mineral interactions. There are diverse range of pore shapes and structures exhibited by CM (Fig. 1.2) and are categorized into capillary pores (tubular with good connectivity) (Fig. 1.2A), ink-bottle pores (narrow necks and wide bodies with poor connectivity) (Fig. 1.2B), plate pores (open in all directions with excellent connectivity) (Fig. 1.2C), stacked pores (also open in all directions with good connectivity) (Fig. 1.2D), slit pores (open at one end with moderate connectivity) (Fig. 1.2E), and mold pores (replicating the shape of single crystals or crystal aggregates, with pyrite mold pores being especially common in shale gas formations) (Fig. 1.2F). The CM are of different types (kaolinite, smectite, illite, and chlorite) and their proportion may vary between different shale gas reservoirs (Keller et al. 1986). Laboratory analysis have observed CM's capacity for adsorption varies depending upon its type. Montmorillonite has the highest adsorption capacity compared to kaolinite, chlorite, and illite [19]. High clay content has been observed to show negative correlation with adsorption in marine shales due to low OM content [3, 47]. Zhang et al. [71] studies the adsorption properties of OM, quartz, and CM in medium-to-high maturity siliceous shale. It was observed that with increase in the kaolinite content the adsorption parameters (pore volume and surface area) also increase. On the other hand, chlorite due to its foliated structure and the ordered arrangement of illite/smectite mixed layers (I/S) resulted in increase of specific surface area. As I/S content decreased and the content of illite increased, the pore volume was observed to increase. At the thermal maturity range of Ro = 1.24–1.34%, OM content was observed to be the dominant factor affecting

Fig. 1.2 Skeleton model of typical clay mineral pores: **A** Capillary pores **B** Ink-bottle-shaped pores **C** Plate-like pores **D** Stacked pores **E** Silt-sized pores **F** Mold pores. The schematic depicts the general pore geometries; however, it is important to note that surface roughness, a key aspect of these pore shapes, plays a significant role. Surface roughness influences fluid dynamics, adsorption potential, and connectivity within the pore network, adding complexity to the transport properties of CM. (Adapted with permission from Xie, W., Chen, S., Vandeginste, V., Yu, Z., Wang, H. and Wang, M., "Review of the Effect of Diagenetic Evolution of Shale Reservoir on the Pore Structure and Adsorption Capacity of Clay Minerals," Energy & Fuels, 2022, 36(9), pp. 4728–4745. Copyright (2022) American Chemical Society)

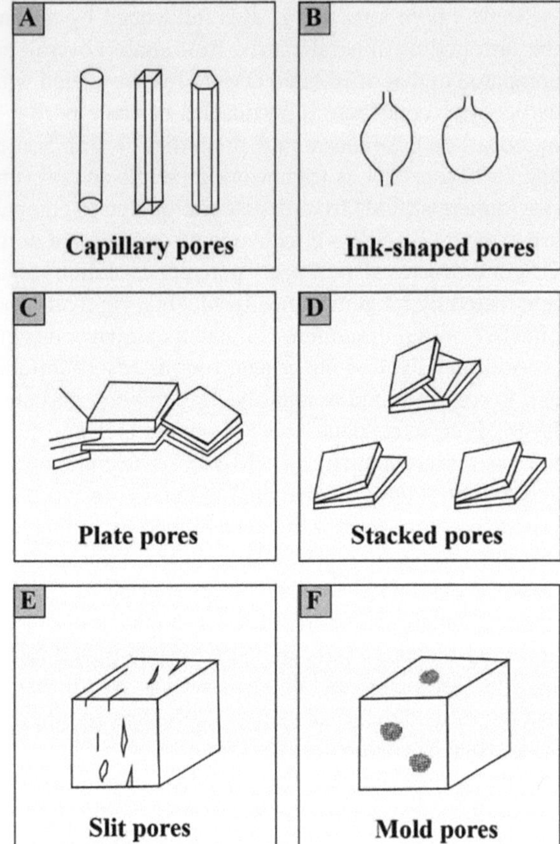

formation of pore. Additionally, the interaction between the siliceous quartz skeleton and overpressure plays a critical role in pore preservation.

1.2.2 Organic Matter, Mineralogical, and Geomechanical Controls on Shale Brittleness

The DOM and mineral content in shale influences another important reservoir parameter: 'brittleness'. The ability of shale reservoirs to develop effective fracture network during fracking depends on its brittleness [44, 52, 58]. Other factors that influence shale brittleness are strength, texture, type of pore fluid, history of diagenesis, reservoir temperature, and in-situ stress conditions [64]. TOC have negative impact on shale brittleness [31]. Thus, high TOC levels can make the rock more ductile. The OM gets easily deformed under stress compared to the surrounding mineral matrix.

As a result, areas with high TOC match up with zones that are less brittle, which may not generate fracture more easily during hydraulic fracturing [64]. On the other hand, zones with lower TOC and more brittle minerals are better spots for developing effective fracture networks. However, this phenomenon is not always straightforward. Shale rock rich in OM show more brittleness when the stress is applied along the bedding plane [31]. This brittleness arises as the organic-rich layers or laminations provide weak surfaces that are prone to fracturing. Thus, OM-rich shales can exhibit both localized ductile behaviour and can also facilitate brittle fracturing depending upon how the stress is applied [68].

Recent advances in techniques like in-situ atomic for microscopy (AFM) combined with optical microscopy enabled researchers to locate OM phases in shales and measure its mechanical properties [61, 69]. Wang et al. [61] used in-situ AFM technique on shales from Wufeng-Longmaxi formation and observed that combining in-situ AFM with nano-indentation provided clearer view of OM mechanical properties than the older stand-alone grid-indentation method. The study also highlighted distinct mechanical response between the OM and mineral phases within the shale sample. Similarly, PeakForce Quantitative Nanomechanical Mapping (PF-QNM) is an AFM technique used for high-resolution mechanical mapping of materials [27, 38]. This method calculates a material's modulus by measuring how an AFM cantilever bends when applying a steady peak force to a very sharp tip smaller than 100 nm in radius [48]. Onwumelu et al. [38] conducted PF-QNM on Bakken shale and observed that Young's modulus of shale decrease with increasing thermal maturity. The average Young's modulus of immature shale was found to be 6 GPa, whereas the early mature and peak mature shale showed average Young's modulus of 5.5 GPa and 3.8 GPa, respectively.

Several studies have shown the interplay between TOC and mineral composition playing a key role in brittleness of shales [28, 49]. Shales with high TOC but with lower amount of quartz show lower brittleness even though the latter generally adds to rock stiffness. Shale samples with over 15 wt.% TOC show weaker tensile strength and less brittleness [49]. Supandi et al. [54] conducted a study analysing the impact of specific clay mineral types on shale mechanical behavior and observed that increase in illite content in shale resulted in decrease of strength, cohesion and friction angle. In contrast, kaolinite was observed to have less impact on the mechanical properties (correlation coefficient <0.3). However, shales with predominant kaolinite composition tend to show higher strength and brittleness index [49]. Kaolinite has higher plasticity index (5–70), making it less prone to plastic deformation compared to illite with plasticity index between 25 and 60 [62].

CM show swelling behavior based on their surface area and cation exchange capacity (CEC). The nature of interaction of CM with water and other fluids is influenced by these two factors impacting the permeability and mechanical properties of the formation. Increased surface area often translates to an increase in overall material permeability, as the more surface area a material has, the more sites there are for fluid interaction to occur, resulting in a dense pore network. At the same time, a greater CEC indicates that mineral's capacity to exchange and adsorb cations, a characteristic that renders CM more prone to expanding once in contact with water

[67]. When comparing the different CM, kaolinite and illite have very different swelling behavior form one another based on their unique characteristics that give them different surface areas and CEC. Kaolinite, with its relatively small surface area usually about 15 m^2/g, and low CEC values of 1–10 meq/100 g show very little reactivity with water. Such limited capacity of water uptake and cation exchange ensures that kaolinite does not expand much in hydrous conditions, making it less cation-exchange prone compared to other CM [20]. In contrast, the larger surface area (30 m^2/g) and high CEC (10–40 meq/100 g) of illite shows higher reactivity when comes in contact with water, making illite more prone to swelling. Formations with higher illite content are more prone to swelling causing negative impacts on permeability and mechanical properties [20]. Thus, shale formations with higher clay content would be ideal for fracturing if the predominant clay mineral is kaolinite.

In addition to TOC and mineral composition, subsurface stress conditions also determine shale brittleness. High-horizontal stress limits fractures growth which results in lower brittleness, whereas lower stress differences increase brittleness [68]. The impact of in-situ stress conditions on rock brittleness is evaluated using conventional and true-triaxial testing [13, 51, 59, 73].

1.3 Depositional Environment Control on OM Enrichment in Shales

Shales deposit in terrestrial, marine and transitional settings [4, 40]. Understanding lithofacies characteristics and their distribution along with the development of depositional models is crucial for understanding the mechanisms that drive OM accumulation, defining reservoir potential, and identifying the most favourable lithofacies for reservoir development [32]. The geochemical features of sediments are controlled by depositional environment recording valuable information about the paleoenvironmental conditions which ultimately helps in interpreting organic rich shales based on lithological and petrographic analysis [2, 43, 72]. The OM enrichment in shales is studied on the basis of two dominant models- (i) the paleoproductivity model: biological productivity in surface water causes preservation of organic-rich sediments [72]; (ii) the preservation model- anoxic or reducing bottom-water condition prevents degradation of OM and promotes its accumulation [36]. Trace elemental and major oxides concentration in shales provide valuable information on the paleoenvironmental condition [34]. Concentration of elements like phosphorus (P), barium (Ba), nickel (Ni), and copper (Cu) can be used as geochemical proxies to determine paleoproductivity [14, 57]. For example, concentration of Ba correlates with high productivity periods as barium sulfate precipitates along with decaying OM. Although paleoproductivity and preservation models are two different mechanisms, the importance of environmental factors in OM enrichment is emphasized by both the models. Ni and Cu are often found enriched in organic-rich shales, thus serve as proxies for biological productivity and redox conditions [42].

OM enrichment is also affected by the change in sedimentary environment the alteration of the flux of terrigenous clastic material and sedimentation rates. Essential nutrients stimulating marine productivity such as planktonic blooms dominated by algae are provided by terrigenous inputs [72]. High rate of sedimentation reduces the exposure of OM from degradation due to oxidation, thereby preserving it [26]. Although if fine-grained detrital material is deposited excessively it can lead to dilution of OM concentrations in sediments [46]. Therefore, balance between sedimentation rates and terrigenous influx can have contrasting impact on OM preservation in shales. A deeper understanding of shale formation and its potential for hydrocarbon source can be enabled by studying the prevailing geochemical indicators. Integration of these insights with lithofacies analysis and depositional models helps researchers to develop predictive models to identify sweet spots and optimization of shale resource development.

References

1. Arif M, Lebedev M, Barifcani A, Iglauer S (2017) Influence of shale-total organic content on CO_2 geo-storage potential. Geophys Res Lett 44(17):8769–8775
2. Borchers SL, Schnetger B, Böning P, Brumsack HJ (2005) Geochemical signatures of the Namibian diatom belt: perennial upwelling and intermittent anoxia. Geochem Geophys Geosystems 6(6)
3. Chen G, Li C, Lu S, Guo T, Wang M, Xue Q, Zhang T, Li Z, Sun Y, Liu J, Jiang S (2021) Critical factors controlling adsorption capacity of shale gas in Wufeng-Longmaxi formation, Sichuan Basin: evidences from both experiments and molecular simulations. J Nat Gas Sci Eng 88:103774
4. Dai J, Zou C, Dong D, Ni Y, Wu W, Gong D, Wang Y, Huang S, Huang J, Fang C, Liu D (2016) Geochemical characteristics of marine and terrestrial shale gas in China. Mar Pet Geol 76:444–463
5. Delle Piane C, Ansari H, Li Z, Mata J, Rickard W, Pini R, Dewhurst DN, Sherwood N (2022) Influence of organic matter type on porosity development in the Wufeng-Longmaxi Shale: a combined microscopy, neutron scattering and physisorption approach. Int J Coal Geol 249:103880
6. Dong M, Gong H, Sang Q, Zhao X, Zhu C (2022) Review of CO_2-kerogen interaction and its effects on enhanced oil recovery and carbon sequestration in shale oil reservoirs. Resour Chem Mater 1(1):93–113
7. Feng G, Li W, Zhu Y, Wang Y, Song Y, Zheng S, Shang F (2023) Scale-dependent fractal properties and geological factors for the pore structure in shale: insights from field emission scanning electron microscopy and fluid intrusion. Energy Fuels 37(21):16519–16535
8. Hackley PC, Cardott BJ (2016) Application of organic petrography in North American shale petroleum systems: a review. Int J Coal Geol 163:8–51
9. Hackley PC, Valentine BJ, Voortman LM, VAN Oosten Slingeland DS, Hatcherian J (2017) Utilization of integrated correlative light and electron microscopy (iCLEM) for imaging sedimentary organic matter. J Microsc 267(3):371–383

10. Hazra B, Singh PK, Sethi C, Pandey JK (2024) Application of optical-electron correlative microscopy for characterization of organic matter. J Geol Soc India 100(10):1385–1394
11. Hazra B, Vishal V, Singh DP (2020) Applicability of low-pressure CO_2 and N_2 adsorption in determining pore attributes of organic-rich shales and coals. Energy Fuels 35(1):456–464
12. Hazra B, Vishal V, Sethi C, Chandra D (2022) Impact of supercritical CO_2 on shale reservoirs and its implication for CO_2 sequestration. Energy Fuels 36(17):9882–9903
13. He BG, Li HP, Meng XR, Dou SY, Yan S, Chen T (2024) Assessment of brittle failure in deep tunnels via true triaxial compression. Tunn Undergr Space Technol 152:105962
14. Horner TJ, Little SH, Conway TM, Farmer JR, Hertzberg JE, Janssen DJ, Lough AJM, McKay JL, Tessin A, Galer SJ, Jaccard SL (2021) Bioactive trace metals and their isotopes as paleo-productivity proxies: an assessment using GEOTRACES-era data. Glob Biogeochem Cycles 35(11):e2020GB006814
15. Hutton A, Bharati S, Robl T (1994) Chemical and petrographic classification of kerogen/macerals. Energy Fuels 8(6):1478–1488
16. International Committee for Coal and Organic Petrology (ICCP) (1998) The new vitrinite classification (ICCP system 1994). Fuel 77:349–358
17. International Committee for Coal and Organic Petrology (ICCP) (2001) The new inertinite classification (ICCP system 1994). Fuel 80:459–471
18. Jacob H (1985) Disperse solid bitumens as an indicator for migration and maturity in prospecting for oil and gas. Erdoel Kohle Erdgas Petrochem 38(8). (Federal Republic of Germany)
19. Ji LQ, Zhang T (2012) Micro-pore characteristics and methane adsorption properties of common clay minerals by electron microscope scanning. Acta Petrolei Sinica 33(2):249
20. Karpiński B, Szkodo M (2015) Clay minerals–mineralogy and phenomenon of clay swelling in oil & gas industry. Adv Mater Sci 15(1):37–55
21. Katz B, Lin F (2014) Lacustrine basin unconventional resource plays: key differences. Mar Pet Geol 56:255–265
22. Keller WD, Reynolds RC, Inoue A (1986) Morphology of clay minerals in the smectite-to-illite conversion series by scanning electron microscopy. Clays Clay Miner 34(2):187–197
23. Knapp LJ, Ardakani OH, Uchida S, Nanjo T, Otomo C, Hattori T (2020) The influence of rigid matrix minerals on organic porosity and pore size in shale reservoirs: upper Devonian Duvernay Formation, Alberta, Canada. Int J Coal Geol 227:103525
24. Krevelen DW (1961) Coal: typology, chemistry, physics, constitution. Elsevier Publishing Company
25. Kuila U (2013) Measurement and interpretation of porosity and pore-size distribution in mudrocks: the whole story of shales. Colorado School of Mines
26. Lash GG, Blood DR (2014) Organic matter accumulation, redox, and diagenetic history of the Marcellus Formation, southwestern Pennsylvania, Appalachian basin. Mar Pet Geol 57:244–263
27. Li C, Ostadhassan M, Gentzis T, Kong L, Carvajal-Ortiz H, Bubach B (2018) Nanomechanical characterization of organic matter in the Bakken formation by microscopy-based method. Mar Pet Geol 96:128–138
28. Li H, Li D, He Q, Sun Q, Zhao X (2024) Controlling mechanism of shale palaeoenvironment on its tensile strength: a case study of Banjiuguan Formation in Micangshan Mountain. Fuel 355:129505
29. Liu B, Mastalerz M, Schieber J (2022) SEM petrography of dispersed organic matter in black shales: a review. Earth Sci Rev 224:103874

30. Liu B, Schieber J, Mastalerz M (2017) Combined SEM and reflected light petrography of organic matter in the New Albany Shale (Devonian-Mississippian) in the Illinois Basin: a perspective on organic pore development with thermal maturation. Int J Coal Geol 184:57–72
31. Liu B, Wang S, Ke X, Fu X, Liu X, Bai Y, Pan Z (2020) Mechanical characteristics and factors controlling brittleness of organic-rich continental shales. J Petrol Sci Eng 194:107464
32. Ma Y, Lu Y, Liu X, Zhai G, Wang Y, Zhang C (2019) Depositional environment and organic matter enrichment of the lower Cambrian Niutitang shale in western Hubei Province, South China. Mar Pet Geol 109:381–393
33. Mastalerz M, Drobniak A, Stankiewicz AB (2018) Origin, properties, and implications of solid bitumen in source-rock reservoirs: a review. Int J Coal Geol 195:14–36
34. McLennan SM, Taylor SR (1991) Sedimentary rocks and crustal evolution: tectonic setting and secular trends. J Geol 99(1):1–21
35. Milliken KL, Zhang T, Chen J, Ni Y (2021) Mineral diagenetic control of expulsion efficiency in organic-rich mudrocks, Bakken Formation (Devonian-Mississippian), Williston Basin, North Dakota, USA. Mar Pet Geol 127:104869
36. Mort H, Jacquat O, Adatte T, Steinmann P, Föllmi K, Matera V, Berner Z, Stüben D (2007) The Cenomanian/Turonian anoxic event at the Bonarelli Level in Italy and Spain: enhanced productivity and/or better preservation? Cretac Res 28(4):597–612
37. Murugesu MP, Joewondo N, Prasad M (2023) Carbon storage capacity of shale formations: mineral control on CO_2 adsorption. Int J Greenh Gas Control 124:103833
38. Onwumelu C, Kolawole O, Nordeng S, Olorode O (2024) Analyzing thermal maturity effect on shale organic matter via PeakForce quantitative nanomechanical mapping. Rock Mech Bull 3(3):100128
39. Peters KE, Cassa MR (1994) Applied source rock geochemistry: chapter 5: part II. Essential elements
40. Piper DZ, Calvert SE (2009) A marine biogeochemical perspective on black shale deposition. Earth Sci Rev 95(1–2):63–96
41. Rine JM, Smart E, Dorsey W, Hooghan K, Dixon M (2013) 12 Comparison of porosity distribution within selected North American Shale units by SEM examination of argon-ion-milled samples.
42. Riquier L, Tribovillard N, Averbuch O, Devleeschouwer X, Riboulleau A (2006) The Late Frasnian Kellwasser horizons of the Harz Mountains (Germany): two oxygen-deficient periods resulting from different mechanisms. Chem Geol 233(1–2):137–155
43. Rowe H, Hughes N, Robinson K (2012) The quantification and application of handheld energy-dispersive x-ray fluorescence (ED-XRF) in mudrock chemostratigraphy and geochemistry. Chem Geol 324:122–131
44. Rybacki E, Meier T, Dresen G (2016) What controls the mechanical properties of shale rocks?–part II: brittleness. J Petrol Sci Eng 144:39–58
45. Safaei-Farouji M, Sethi C, Hazra B, Gao Y, Fattahi E, Vishal V, Pandey JK, Ostadhassan M (2024) CO_2 storage potential of coaly shales of the Barakar formation in the Rajmahal basin, India. Energy Fuels 38(15):14173–14187
46. Sageman BB, Murphy AE, Werne JP, Ver Straeten CA, Hollander DJ, Lyons TW (2003) A tale of shales: the relative roles of production, decomposition, and dilution in the accumulation of organic-rich strata, Middle-Upper Devonian, Appalachian basin. Chem Geol 195(1–4):229–273
47. Sander R, Pan Z, Connell LD, Camilleri M, Grigore M, Yang Y (2018) Controls on methane sorption capacity of Mesoproterozoic gas shales from the Beetaloo Sub-basin, Australia and global shales. Int J Coal Geol 199:65–90

48. Sedin DL, Rowlen KL (2001) Influence of tip size on AFM roughness measurements. Appl Surf Sci 182(1–2):40–48
49. Sethi C, Hazra B, Ostadhassan M, Motra HB, Dutta A, Pandey JK, Kumar S (2024) Depositional environmental controls on mechanical stratigraphy of Barakar shales in Rajmahal basin, India. Int J Coal Geol 285:104477
50. Sethi C, Mastalerz M, Hower JC, Hazra B, Singh AK, Vishal V (2023) Using optical-electron correlative microscopy for shales of contrasting thermal maturity. Int J Coal Geol 274:104273
51. Shen R, Wang X, Li H, Gu Z, Liu W (2024) Brittleness characteristics and damage evolution of coal under true triaxial loading based on the energy principle. Nat Resour Res 33(1):421–434
52. Sone H, Zoback MD (2013) Mechanical properties of shale-gas reservoir rocks—part 1: static and dynamic elastic properties and anisotropy. Geophysics 78(5):D381–D392
53. Suárez-Ruiz I, Flores D, Mendonça Filho JG, Hackley PC (2012) Review and update of the applications of organic petrology: part 1, geological applications. Int J Coal Geol 99:54–112
54. Supandi S, Zakaria Z, Sukiyah E, Sudradjat A (2019) The influence of kaolinite-illite toward mechanical properties of claystone. Open Geosci 11(1):440–446
55. Thommes M, Kaneko K, Neimark AV, Olivier JP, Rodriguez-Reinoso F, Rouquerol J, Sing KS (2015) Physisorption of gases, with special reference to the evaluation of surface area and pore size distribution (IUPAC technical report). Pure Appl Chem 87(9–10):1051–1069
56. Tissot BP, Welte DH (1984) Kerogen: composition and classification. In: Petroleum formation and occurrence, pp 131–159
57. Tribovillard N, Algeo TJ, Lyons T, Riboulleau A (2006) Trace metals as paleoredox and paleoproductivity proxies: an update. Chem Geol 232(1–2):12–32
58. Tuzingila RM, Kong L, Kasongo RK (2024) A review on experimental techniques and their applications in the effects of mineral content on geomechanical properties of reservoir shale rock. Rock Mech Bull 100110
59. Vachaparampil A, Ghassemi A (2017) Failure characteristics of three shales under true-triaxial compression. Int J Rock Mech Min Sci 100:151–159
60. Vandenbroucke M, Largeau C (2007) Kerogen origin, evolution and structure. Org Geochem 38(5):719–833
61. Wang J, Dziadkowiec J, Liu Y, Jiang W, Zheng Y, Xiong Y, Renard F (2024) Combining atomic force microscopy and nanoindentation helps characterizing in-situ mechanical properties of organic matter in shale. Int J Coal Geol 281:104406
62. Wilson MJ, Wilson L (2014) Clay mineralogy and shale instability: an alternative conceptual analysis. Clay Miner 49(2):127–145
63. Yang G, Zeng J, Qiao J, Liu Y, Liu S, Jia W, Cao W, Wang C, Geng F, Wei W (2022) Role of organic matter, mineral, and rock fabric in the full-scale pore and fracture network development in the mixed lacustrine source rock system. Energy Fuels 36(15):8161–8179
64. Yasin Q, Du Q, Sohail GM, Ismail A (2017) Impact of organic contents and brittleness indices to differentiate the brittle-ductile transitional zone in shale gas reservoir. Geosci J 21:779–789
65. Zargari S, Canter KL, Prasad M (2015) Porosity evolution in oil-prone source rocks. Fuel 153:110–117
66. Zhan H, Wu S, Bao R, Ge L, Zhao K (2015) Qualitative identification of crude oils from different oil fields using terahertz time-domain spectroscopy. Fuel 143:189–193
67. Zhang C, Pathegama Gamage R, Perera MSA, Zhao J (2017) Characteristics of clay-abundant shale formations: use of CO_2 for production enhancement. Energies 10(11):1887
68. Zhang D, Ranjith PG, Perera MSA (2016) The brittleness indices used in rock mechanics and their application in shale hydraulic fracturing: a review. J Petrol Sci Eng 143:158–170
69. Zhang R, Cao J, Hu W, Zuo Z, Yao S, Xiang B, Ma W, He D (2022) Nanomechanical characterization of organic-matter maturity by atomic force microscopy (AFM). Int J Coal Geol 261:104094

70. Zhang T, Ellis GS, Ruppel SC, Milliken K, Yang R (2012) Effect of organic-matter type and thermal maturity on methane adsorption in shale-gas systems. Org Geochem 47:120–131
71. Zhang W, Zhou S, Yu Z, Liu X, Wang S, Miao H, Liu D, Tian J, Wang H (2025) Controls of clay mineral transformation and organic matter on pore networks of the Paleogene lacustrine shale oil system in the Yitong Basin, NE China. J Asian Earth Sci 280:106469
72. Zhao J, Jin Z, Jin Z, Geng Y, Wen X, Yan C (2016) Applying sedimentary geochemical proxies for paleoenvironment interpretation of organic-rich shale deposition in the Sichuan Basin, China. Int J Coal Geol 163:52–71
73. Zhou X, Liu H, Guo Y, Wang L, Hou Z, Deng P (2019) An evaluation method of brittleness characteristics of shale based on the unloading experiment. Energies 12(9):1779

Chapter 2
Advanced Techniques for Geochemical Characterization

Abstract The Rock–Eval (RE) technique provides a rapid and reliable way to geochemical screening of organic-rich shales. It assesses the amount of organic carbon present, and provides indications of the type of organic matter present (oil-prone and/or gas-prone kerogen) from the hydrogen index (HI), and indirect measures of thermal maturity from the temperature profiles of its S2 peak. Recent studies have revealed that additional compositional and thermal maturity data can be extracted from RE's S3 and S4 peaks, respectively. It has been observed that shale samples need to be carefully prepared applying limits to their sample weight and particle size to obtain consistent/repeatable results. Using pyrolysis analysis at three heating rates enables the reaction kinetics to be determined for shales and/or their component kerogens. To achieve this effectively, the Arrhenius equation is used to fit the S2 peaks of the three heating rates in parallel with an optimizer allowing both activation energy (E) and pre-exponential factor (A) to vary for a set of approximately eleven first-order reactions. Each reactions E and A values used to achieve the S2 peak fit with low error are weighted in accordance to their fractional contribution. The weighted average and weighted standard deviations of the E and A values of the fitted reactions tend to closely follow an empirically observed E versus lnA trend. The E and A values can also be used to quantify the hydrocarbon generation fraction versus temperature, a key PSM requirement. A case study of shale samples from the Damodar Valley Permian basins of northeast India illustrates how shale reaction kinetic can be derived from the RE S2 peaks and interpreted. X-ray diffraction (XRD), X-ray fluorescence (XRF), and inductively coupled plasma mass spectrometry (ICP-MS) provide details of the mineralogical and elemental compositions of shales. XRF scanning of core samples with hand-held devices provide rapid and detailed geochemical characterization. Machine learning models including cluster analysis can assist in the interpretation of large XRF datasets. ICP-MS analysis provides fine-scale elemental details, which are useful for defining geochemical pathways through a shales pore network, mineralogical transformations and environmental risks of chemical pollution from fracture-stimulation fluid flow back.

C. Sethi et al., *Recent Advances in Shale Characterization Based on Laboratory Geochemistry and Geomechanics Techniques*,
SpringerBriefs in Petroleum Geoscience & Engineering,
https://doi.org/10.1007/978-3-032-03961-3_2

Keywords Rock–Eval pyrolysis · Geochemical screening · 1D to 3D Petroleum system modelling (PSM) · Reaction kinetics · Arrhenius equation to fit S2 pyrolysis peak · Multi-heating rate pyrolysis · X-ray diffraction (XRD) · X-ray fluorescence (XRF) · Inductively coupled plasma mass spectrometry (ICP-MS) · Geochemical proxies

2.1 Introduction

Geochemical investigations, comprising of both inorganic and organic components, have been commonly applied by researchers to comprehensively evaluate shale systems in many regions. Amongst the first, and probably the most essential component, particularly when assessing hydrocarbon plays such as shales, is 'geochemical screening' using programmed pyrolysis/oxidation (such as RE) for evaluating the nature, amount, thermal maturity of sedimentary organic matter (OM) [9, 12]. Programmed pyrolysis typically involves heating of rock sample in an oven filled with an inert gas, usually nitrogen or helium. The oven follows a specific heating schedule, which has changed over the years between older systems like RE II and the newer RE VI. As the sample heats up, it starts releasing hydrocarbons, which are then picked up by a flame ionization detector. At the same time, the released CO_2 is measured using an infrared detector. These measurements help quantification of several basic parameters (Table 2.1) [4].

2.2 Recent Progress on Application of Rock–Eval for Shale Systems

RE pyrolysis has undergone considerable development in recent years, improving its use as a tool in assessment of shale systems. All of these developments have made characterizing hydrocarbon generation, expulsion, retention, and thermal maturity in shales more reliable. Hazra et al. [11] used the RE technique to conduct source-rock geochemical analysis of Permian shales of Rajmahal basin (RB) and observed that RE S4Tpeak can be a useful alternative for reliably assessing thermal maturity in organic-rich shales. They also observed that widely accepted maturity proxies such as T_{max} and vitrinite reflectance (VRo) can often provide ambiguous or conflicting results, particularly in rocks at higher stages of thermal maturity. This is because T_{max} frequently flattens out or takes an erratic form under such conditions owing to suppression effects and/or the impacts of mixed kerogen types. Therefore, their study focused on the behavior of more thermally resistant OM under oxidation.

In contrast to the pyrolysis S2 signal, which lacks reliability when analysing thermally altered or overmature samples as a result of diminished hydrocarbon yields and noise in FID response, the $S4CO_2$ oxidation curve maintained a symmetric profile

Table 2.1 Rock–Eval parameters

Parameters	Unit (Name)	Definition
S1	mg HC/g rock	Measure of the amount of free hydrocarbons (HC) present in the rock. The free HC are those thermally released at 300 °C for 3 min
S2	mg HC/g rock	Measure of HC formed by cracking kerogen from 300 to 650 °C at 25 °C/min
T_{max}	°C	Maximum temperature of HC generation by cracking of kerogen (S2)
S3	mg CO_2/g rock	Measure of CO_2 generated from oxygeanated functional groups in kerogen between 300 to 400 °C i.e., CO_2 from organic sources
S3′	mg CO_2/g rock	Measure of CO_2 generated from minerals above 400 °C i.e., CO_2 from mineral sources
PC	Wt%/ Pyrolysable org. Carbon	Amount of C pyrolysed contained in organic phases
RC	Wt%/ Residual org. C	Measure of the residual organic carbon in a rock with no HC generation potential
TOC	Wt%/Total org. C	Measures the rock's organic richness
HI	mg HC/g TOC	Amount of HC released on pyrolysis (S2) normalized to TOC
OI	mg CO_2/g TOC	Amount of CO_2 released on pyrolysis (S3) normalized to TOC

and generated a robustly interpretable IR response. Unusually poor S2 curve was observed for the heat affected shale with T_{max} of 509 °C. Conversely, the S4Tpeak was observed to be 643 °C, offering a much more stable and chemically meaningful indicator of its high thermal maturity.

Similarly, asymmetrical S2 curve was also observed for post-mature shale but its oxidation phase S4 curve was well resolved and symmetrical. Symmetrical S4 oxidation curve was also observed for less mature shales displaying the robustness of the parameter across different thermal maturity levels. Overall, this case study demonstrated that S4Tpeak supplements the information delivered by routine RE analysis and also increases usefulness of the method in characterizing highly mature rocks for which standard proxies may start to fail. Therefore, S4Tpeak can be used as a powerful tool for basin modelling, resource evaluation, and elucidating hydrocarbon generation processes in mature shale systems.

Hazra et al. [10] identified certain mineral phases in shales using the RE technique. Presence of siderite in the shale samples was observed through X-ray diffraction (XRD) and petrographic analysis. Higher S3′ values was observed for siderite rich samples indicating carbonate, and not OM, as the largest source of CO_2 release during pyrolysis. Therefore, unusually high OI was observed for these samples. This was not the result of increased oxidation of the OM, it was because the CO_2

released from the carbonates, particularly from siderite, was erroneously attributed to organic sources when S3 was not corrected. The study also found decrease in HI with increasing siderite content. This occurred as a result of siderite's contribution of CO_2 interfering with the pyrolysis results and increasing the TOC values without increasing the hydrocarbon potential.

Shale samples blended with limestone was also assessed in that study. Similar to the observation made for siderite, the TOC of the shale samples decreased with increasing limestone content. Also, shale sample mixed with limestone resulted in increasing OI however the magnitude was less compared to shale sample mixed with siderite. The values of OI did not show any significant change for samples with 25–50% limestone mixture. Their study highlighted breakdown of siderite begins at temperatures below 400 °C which is within the early phase of the RE pyrolysis heating cycle. Thus, the release of CO_2 with the breakdown of siderite is recorded during the initial phase of gas evolution. If this is not taken into consideration, it can be misattributed to the oxidation of OM and overestimate the OI, possibly affecting the accurate evaluation of the kerogen quality.

Similar trends were observed by Sethi and Hazra [27] on a suite of Permian shales from RB, India. In their studied range of samples, the maximum OI was found in sample RM-15X2, which displayed the greatest S3′ value. Likewise, sample RM-15, with a high OI of 22 had a very high value of S3′. Conversely, sample RM-6 with an OI of 31 was accompanied by a lower value of S3′. XRD confirmed the presence of siderite in RM-15 and RM-15X and that no siderite was present in RM-6. This was interpreted to suggest that, among these samples, both OI and S3 values were influenced by two primary factors: (a) the presence of siderite (as evidenced by the high S3 and S3′ values in RM-15 and RM-15X2); and (b) the presence of oxygenated organic compounds, which likely account for the elevated OI in siderite-free sample RM-6.

Figure 2.1 displays the S3 and S3′ curves of the shale samples. Reaction profiles for samples RM-15 (Fig. 2.1A) and RM-15X2 (Fig. 2.1B) showing pronounced release of CO_2 at temperature above 400 °C, indicative of siderite decomposition during pyrolysis. Samples RM-6 (Fig. 2.1C) and RM-8X2 (Fig. 2.1D) show noisy S3′ signals indicative of siderite's absence. The IR CO_2 signal during pyrolysis is similarly erratic for these two samples. The lower reliable limit for the RE 6 IR detector is 0.2 mV, below which the CO_2 IR signal falls. This value was observed to fall below this threshold for RM-6 sample, while in RM-8X2 sample it surpasses it at 0.225 mV. Even with its low S3 value, RM-6 exhibits a high OI, which is due to its low TOC content (2.1 wt%). A lower TOC may create an artificial inflation of the OI. This is due to the fact that a higher TOC value would have likely resulted in a lower OI for this sample. These studies highlight the benefit of the RE methodology for identification of carbonate phases via CO_2 evolution patterns.

Sing et al. [32] found that both the particle crush size and sample weight had a major impact on RE pyrolysis parameters. Overall, finer particle sizes led to consistently higher S2 and PC values for all shale types. This was due to the increased surface area of OM in smaller sizes particles, allowing reactions to proceed to a fuller reaction during pyrolysis. There was no discernible trend in T_{max} values. Sample

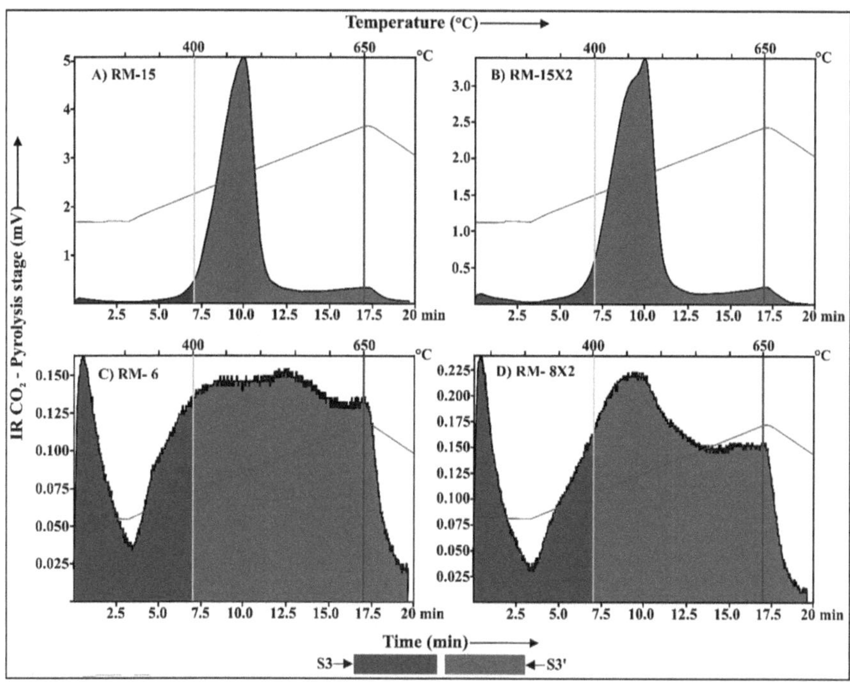

Fig. 2.1 S3 and S3′ curves generated during the pyrolysis phase. The section of the curve below 400 °C corresponds to the S3 region, while the portion extending from 400 °C to the final pyrolysis temperature is designated as the S3′ region. (Reproduced from Sethi, C. and Hazra, B., "Source rock properties of Permian shales from Rajmahal Basin, India," *Arabian Journal of Geosciences*, 15(13), p. 1219, 2022, with permission from Springer Nature. License Number: 6030330696419)

weight was another important factor. The magnitudes of S2 signal became larger with greater sample weights. This trend was consistent regardless of shale maturity. Significant reduction in TOC and RC was detected with increasing sample weight for mature Permian shales, indicating that CO_2 may have been under-detected during oxidation when larger sample weights were employed. For immature shales such as the Palaeocene lignite samples, increasing the sample weight had no effect on TOC because the OM burned off at lower temperatures. The shape of these S2 and S4 curves was also observed to be affected by particle size and weight. Specifically, mature shales showed a consisted rightward shift of the $S4CO_2$ curve with increased sample weight suggesting progressive oxidation inhibition.

The study emphasized the predominant effect of OM type and thermal maturity on the S4Tpeak. High maturity, type III-IV kerogen rich shales produced oxidation curves that extended past the conventional boundary (650 °C) and resulted in likely underestimation of TOC values at heavier sample weights. Sethi et al. [29] observed significant impact of sample weight on S4Tpeak parameter of RE. Figure 2.2 displays the evolution of $S4CO_2$ profiles of three distinct shale samples analysed at two different sample weights. In particular, the $S4CO_2$ curve of RB5 exhibits a clear hinge with the presence of two sub-peaks. One of these, shown on the upper left part of the curve, is spikey (see Fig. 2.2A and B). At the lowest sample weights this low-temperature spiky sub-peak yields a stronger signal than the sub-peak at higher temperature. Consequently, the S4TPeak of the low-weight sample is seen at 395 °C (Fig. 2.2A), while for the heavier weight, it moves to 425 °C (Fig. 2.2B). The RN and JB shales have significantly higher S4TPeaks (554 °C and 568 °C, respectively) in all sample weights indicating their relatively higher thermal maturity.

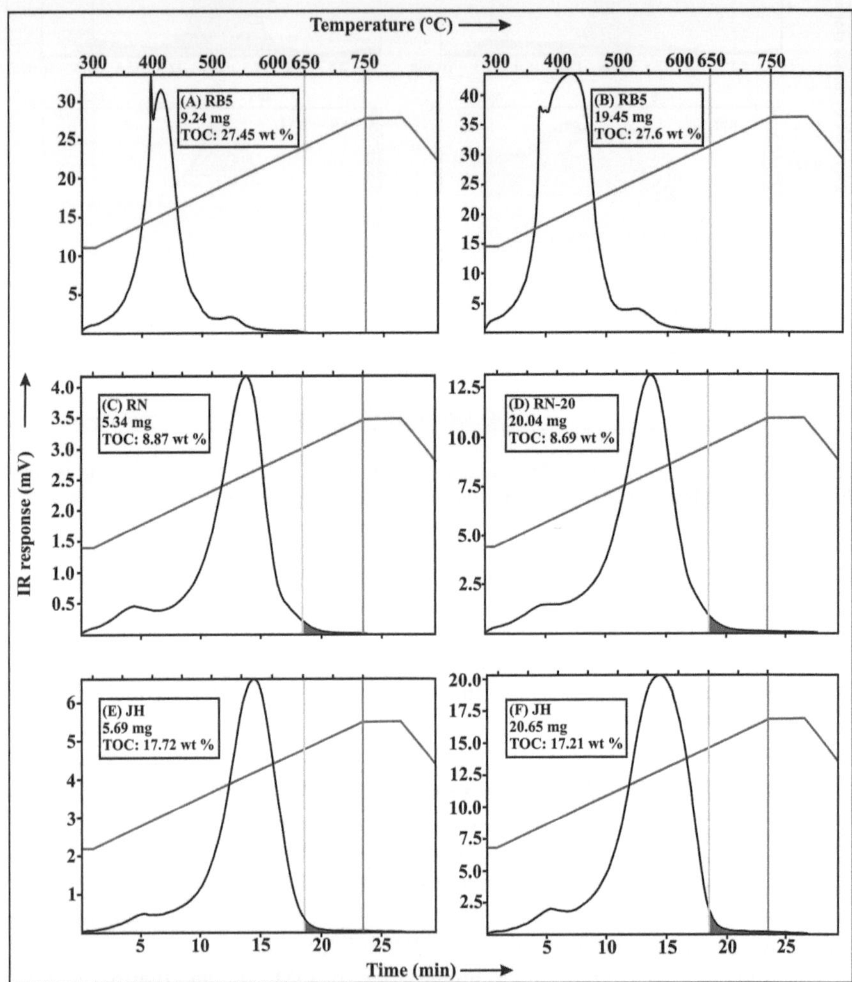

Fig. 2.2 S4CO$_2$ profiles of shale samples measured at two different sample weights. The grey-shaded areas indicate segments of the organic CO$_2$ release (S4CO$_2$) that are not fully accounted for in the total organic carbon (TOC) estimation. (Reproduced from Sethi, C., Hazra, B., Wood, D.A. and Singh, A.K., "Experimental protocols to determine reliable organic geochemistry and geomechanical screening criteria for shales," *Journal of Earth System Science*, 132(2), p. 45, 2023, with permission from Springer Nature. License Number: 6030340384625)

Closer inspection of the S4CO$_2$ profiles uncover clear features indicative of the thermal maturity stages of these shales. For RB5 from the RB, the CO$_2$ evolution curves at both sample weights fall completely to the left of the S4-S5 boundary (shown as a yellow line in Fig. 2.2A–F). Consequently, even at a higher sample weight with RB5 no reduction in TOC is seen (Fig. 2.2B). For the more thermally evolved RN and JB shales, much of their S4 CO$_2$ signals (marked in grey in Fig. 2.2C–F) protrude past the S4-S5 boundary even at low sample weights ($\leq$5 mg). This extension is explained by the occurrence of greater amounts of aromatized OM in these samples, a defining characteristic of deeper thermal maturity. This OM is resistant to full oxidation under temperatures of 650 °C, leading some portion of the S4 curve to extend beyond this temperature. These large areas of influence are not reflected in the RC contribution to the TOC calculation, likely leading to a significant underestimation of TOC. Because these samples are devoid of carbonates, all CO$_2$ released during subsequent oxidation comes exclusively from decomposition of OM.

Liao et al. [19] through grain-based RE analysis of lacustrine shales indicated that the temperature of maximum hydrocarbon generation was uniformly higher for whole-rock samples than for the separated kerogen. This pattern even pointed to the idea that the mineral matrix acts as a barrier to hydrocarbon release. Juxtaposing minerals pyrolysis against their kerogen removed from grains permitted researchers to calculate the proportion of hydrocarbon that could potentially be preserved in a given rock at varying maturities. All of the shales didn't act the same in how they expelled hydrocarbons vs retained them. For example, Nenjiang shale was expected to hold more hydrocarbons but fail to discharge them effectively, and the Shahejie shale samples were opposite. Differences in pore structure were very important drivers of these differences. Overall, these grain-based RE method in combination have opened up a fresh door of possibility into pyrolysis analytical techniques, offering a more accurate representation of the behavior of source rocks under heat.

Li et al. [18] focused on the vertical and lateral variation of different geochemical parameters of lacustrine oil shales and connected these variations to changes in the depositional environment, extent of water column stratification, and primary productivity during deposition. Their results confirmed that the investigated shales were primarily Type I kerogen dominant, as evidenced by elevated high HI and OI values, which suggests excellent oil-generating capacity. A strong positive correlation (R^2 > 0.95) was observed between S2 and TOC, indicating high TOC content was a consequence of the presence of oil-prone OM, rather than inert or degraded material. The relatively homogenous distribution of HI and TOC in the sampled shales, as well as minimal variations in T_{max} between the different layers, indicated that depositional conditions were very consistent while the oil shale formed. The organic richness and quality were attributed to high primary productivity in a deep, stratified lake environment, coupled with rapid burial under anoxic conditions at the bottom of the lake, which aided in the prevention of degradation.

2.3 Kinetics of Hydrocarbon Generation

Analysis of the reaction kinetics of organic-rich shales and their component kerogens provides essential information for petroleum system modelling (PSM). Such modelling is now routinely conducted at 1-dimensional (1D), two-dimensional (2D) and three-dimensional (3D) scales, and serves as an effective petroleum exploration and development tool. PSM should incorporate the burial history (depth versus time) of sedimentary sequences, incorporating subsidence and erosion episodes and compaction, the evolution of subsurface formation temperatures as burial progresses, considering dynamic geothermal gradients and formation thermal conductivities, and whole-rock reaction kinetics of the organic-rich formations (shales and coals) present [35]. 1D PSM is conducted at a single location (e.g., a wellbore or an outcrop), 2D PSM is conducted along a cross-section (e.g., correlating between several wellbores), and 3D PSM is conducted basin-wide, typically including data from multiple wellbores and seismic data [3, 30]. To achieve this successfully requires rigorous and detailed characterization of the organic-rich shale formations involved [43]. PSM has been conducted for several decades [33, 37] and its methods progressively refined and improved [38]. However, the majority of PSM studies reported to date either use over-simplistic, generic assumptions for kerogen kinetics based on broadly categorized kerogen types (i.e., Type I, II, III, and IV), or calculate the kerogen kinetics applying geologically unrealistic assumptions to the whole-rock or kerogen pyrolysis experimental data [41].

Applying the Arrhenius equation for first-order reactions (Arrhenius [1], Eq. (2.1)) to fit S2 pyrolysis curves to provide insight to kerogen and whole-rock samples was shown to be effective in the 1970's [33, 34] and has subsequently been widely used for that purpose.

$$k_{Arrhenius} = Ae^{(-E/RT)} \tag{2.1}$$

where $k_{Arrhenius}$ is the first-order reaction rate constant, A is the pre-exponential (frequency) factor expressed in per time units (e.g., 1/minutes for laboratory tests; 1/millions of years for burial-history modelling), E is the activation energy (kJ/mol), and T is absolute temperature ($^{\circ}$K). The reaction kinetics can be expressed in terms of E and A values but to model the progress of a reaction in small temperature and time intervals until it is completed the Arrhenius equations needs to be integrated. For pyrolysis S2 curves generated at constant heating rates throughout the reaction process relatively simple integrals can be applied. However, for geological modelling with the integral needs to have the flexibility to deal with variable heating rates applied over geological time responding to substantial variations in burial/uplift rates and geothermal gradients. The time–temperature index (TTI_{ARR}) (Wood [37, 40], Eq. (2.2)) achieves this effectively and can be used to model normalised (0–1 scale) S2 curve data in one degree temperature intervals.

$$TTI_{ARR}(t_n tot_{n+1}) \approx \frac{A}{qn} \frac{RT_{n+1}^2}{E + 2RT_{n+1}} e^{-E/RT_{n+1}} \tag{2.2}$$

where q_n is the heating rate (°C/my) at time t_n. TTI_{ARR} is typically summed over all the time increments evaluated to provide a cumulative TTI_{ARR} $\left(\sum TTI_{ARR}\right)$ thermal maturity indicator. $\sum TTI_{ARR}$ can be calibrated with measured thermal maturity indicators, such as vitrinite reflectance (Ro %) that are routinely measured for organic-rich shale samples.

The transformation fraction (TF) of the of a reaction is then expressed at time t by Eq. (2.3).

$$TF_t = 1 - e^{-\sum TTI_{ARR}} \tag{2.3}$$

TF is expressed on a scale of 0 (reaction yet to commence) to 1 (reaction completed) and can be monitored for each °C increment and reported for the full reaction temperature range. It is *TF* that is used in PSM to identify the timing and depths of petroleum generation. It is necessary to calculate TF by summing the weighted contributions of a series of parallel reactions, each with their own E and A values, to accurately fit whole-rock shale and/or kerogen S2 curves. Once *TF* has been calculated it can also be used estimate the degree of OM conversion to petroleum in terms of the declining hydrogen index (HI; Eq. (2.4)) of the shale as it passes through the thermal maturation process. This is typically calculated with Eq. (2.4), expressing HI in normalized terms (0–1).

$$HI_{tn} = (1 - TF_{tn}) * HI_{t0} \tag{2.4}$$

PSM can usefully display calculated Ro% (inverted from $\sum TTI_{tn}$), TF_{tn}, HI_{tn} and temperature evolutionary trends superimposed on burial history cross sections. Such displays are useful for determining the timing at which shale intervals of interest reach early-stage, peak, and late-stage thermal maturity. Such information is useful for determining whether a trap was present when petroleum generation began and to estimate the timing of petroleum migration into other potential reservoir.

The precise fitting of RE S2 curves generated from shale or kerogen samples requires multiple reactions to be considered collectively as a kinetic-reaction distribution and is not a straightforward process with a unique solution [36]. To obtain verifiable reaction kinetics from pyrolysis experiments it is necessary to repeat those experiments at multiple (three or more) heating rates and fit the S2 curves from those multiple trials collectively [24]. Attempting to invert kerogen kinetics from single heating-rate S2 curves of multiple samples of different thermal maturities [5] is fraught with uncertainties and requires speculative assumptions regarding the appropriate reaction-kinetic variables to apply.

However, a fundamental disagreement exists in the literature as to how the kinetics of those multiple reactions should be calculated. Peters et al. [25] advocate using a universal value of A ($A = 1 \times 10^{14}$/sec for their Basin%Ro method; updated from 1×10^{13}/sec for their Easy%Ro method developed in the 1990s). Wood [39] demonstrated that by using an optimizer and letting E and A values of multiple reactions vary independently, the weighted average E and A values of those reactions followed the kinetic trend established for known hydrocarbon materials (1988).

For multiple published shale and kerogen samples assessed the weighted average E and A values range substantially along that trend (Fig. 2.3). This suggests that applying a universal A value for kerogen kinetic pyrolysis fits involving multiple reactions, or for generic kerogen types, does not reflect the diversity of the reaction kinetics involved. That diversity is a consequence of shales containing a diverse mixture of kerogen compositions (related to depositional and burial conditions) plus bitumen and reaction residues generated at early, peak, and late stages of thermal maturation [22]. Bitumen is a key intermediate kerogen reaction product and is present to a degree in shale samples [21], particularly those that have reached or exceeded the early stages of thermal maturity. Indeed, individual shale formations can contain several kerogen compositions. Moreover, each generic kerogen type (I, II, III and IV) can display a substantial range of reaction kinetics related to differences in their detailed compositions. The weighted average kinetics of each whole-rock shale sample has a substantial impact on the timing and temperature of its TF_t distribution. Hence, restricting the A values used in S2 curve modelling makes no sense.

For PSM purposes, it is recommended to use kerogen reaction kinetics established with S2 pyrolysis curves generated by thermally immature organic-rich shale samples (Ro <= 0.5%). The reasons for this preference are:

- Such samples have not yet been subjected to temperatures high enough to stimulate extensive thermogenic reactions. Hence, their kerogen is relatively pristine, as are their TOC and HI values, which are useful for assessing the volumes of petroleum they are capable of generating as thermal maturity progresses.
- Early- and peak-mature shales are prone to have bitumen and reaction residues present that broaden the S2 pyrogram curves and the E-A range of the calculated reaction kinetics distributions.
- Late- and post-mature shales are prone to contain natural gases and cracked bitumen products that broaden the S2 pyrogram curves and the E-A range of the calculated reaction kinetics distributions.
- The nano- and meso-pore-size distribution (pore volumes and surface areas) of late- and post-mature shales increase substantially compared to immature/early mature shales. These small pores tend to contain generated natural gases that are released during pyrolysis but the temperature of that release is controlled by pore connectivity as well as kinetics.

The generation of petroleum from organic-rich shales is not only associated with thermogenesis. During burial and the process of shale and kerogen formation, the organic-rich materials present are typically subjected to biogenic and diagenetic reactions before thermogenesis begins. Biogenic gas (methane and hydrogen) can be generated through a series of microbe-induced, first-order kinetic reactions at low maturity, and is often accumulated and preserved in shallow reservoir [26]. Hence, even in an immature state (Ro ~0.3 to 0.5%) a shale's kerogens are typically not "pristine" but contain some traces of the products and residues of biogenic/

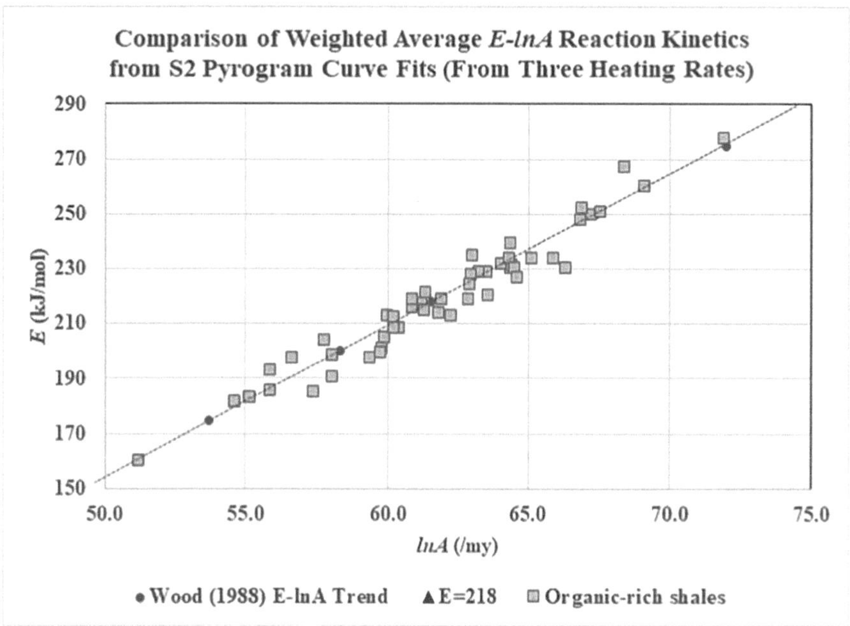

Fig. 2.3 Activation energy (E) versus pre-exponential factor (lnA) multi-heating rate RE S2 curve pyrolysis optimum fits (weighted averages) for multiple shale samples from organic-rich shale formations distributed around the world. The established E-lnA trend [37] is displayed for comparison. (Reproduced from Wood, D.A., "Kerogen Kinetic Distributions and Simulations Provide Insights into Petroleum Transformation Fraction (TF) Profiles of Organic-Rich Shales," *Journal of Earth Science*, 35(3), pp. 747–757, 2024, with permission from Springer Nature. License Number: 6038820940028)

diagenetic reactions. This is particularly the case with shale samples collected from surface outcrops, which are more likely to have been subjected to recent biogenic degradation, whether they are thermally immature or mature. Pyrolysis tests on such samples generate S2 pyrograms that are influenced by contributions from second-order reactions related to these biogenic/diagenetic products and residues, as well as the predominant thermogenic first-order reactions affecting the kerogen itself. The involvement of such reactions explains why even the reaction kinetic distributions derived from whole-rock immature organic-rich shales, or their separated kerogens, do indicate that multiple reactions with a wide range of *E-A* values are involved.

Nevertheless, reaction kinetic distributions determined for shale samples covering a range of thermal maturities can provide useful formation-characterization insights. For instance, the form of the calculated kinetic distribution, and shape of the S2 pyrogram curves [40] are, to an extent, related to certain pore-size distribution characteristics (e.g., pore volume, surface area, and fractal dimension). Hence, kerogen kinetic distributions can be used to identify zones within shale formations of specific thermal maturity that have micro-porosity distributions sufficiently developed to enhance their gas-storage capacities. Such information is useful for potential natural gas productivity and resource recovery calculations. It is also useful for high-grading zones within a shale formation that have suitably high gas-storage capacities to be considered as potential carbon dioxide (CO_2) storage reservoirs.

Three shale samples from the Damodar Valley Permian basins of northeast India have been subjected to multi-heating rate RE pyrolysis to explore the impacts of thermal maturity on organic-rich shales. The pyrolysis S2 curves and pore properties of shale samples from Permian basins of the Damodar Valley have been described previously [31, 42]. The three samples evaluated here, SMR-2B (early thermal maturity, Ro = 0.5%; Mand-Raigarh Basin), MJNK-14 (peak thermal maturity, Ro = 0.77%; North Karanpura Basin) and J-34 (late thermal maturity, Ro = 1.1%, Jharia Basin) were each subjected to RE 6 pyrolysis at heating rates of 25 °C/min, 15 °C/min, and 5 °C/min. Figure 2.4 displays the normalized S2 pyrogram and TF curves with the optimized curve fits superimposed for these three shale samples. The kinetic distributions reported for these samples are those used to generate the optimized curve fits.

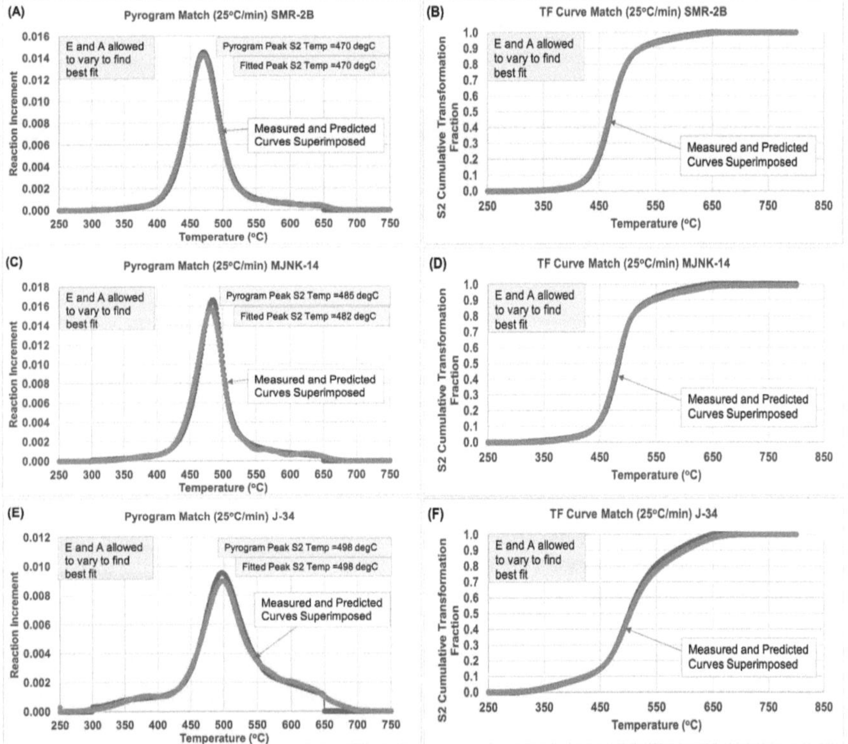

Fig. 2.4 RE normalized S2 curves (left) and transformation fractions (TF) curves (right) for three studied shale samples fitted to the 25 °C/min heating ramp curves. **A** and **B** SMR-2B (early mature); **C** and **D** MJNK-14 (peak mature); and **E** and **F** J-34 (late mature). The optimum kinetic fits are superimposed on the recorded curves in the graphics displayed

Table 2.2 Summaries of RE S2 curve kinetic distribution derived from pyrogram curve fits from three heating ramps for the three studied shale samples. "Wt Avg" refers to weighted average and "Wt SD" refers to weighted average standard deviation

Kinetic distribution summary based on curve fits to RE S2 curves from three heating ramps for each sample			
Sample identifier:	SMR-2B	MJNK-14	J-34
Vitrinite Reflectance (Ro%)	0.50	0.77	1.10
Wt Avg E (kJ/mol)	246.96	248.15	229.69
Wt SD E (kJ/mol)	46.15	42.57	65.32
Wt Avg lnA (/min)	39.28	38.91	34.76
Wt SD lnA (/min)	6.14	7.36	9.26
Wt Avg lnA (/my)	66.27	65.89	61.74
Wt SD lnA (/my)	6.14	7.36	9.26
Range of E values (kJ/mol)	197	185	241
Curve fitting error (ms*1000)	0.3791	0.8374	0.6804

Samples SMR-2B and MJNK-14 have both generated reasonably symmetrical S2 pyrograms. However, the peak area of the pyrogram for MJNK-14 is substantially narrower than that of SMR-2B, although both samples display evidence of a small right-side shoulder in their S2 curves. On the other hand, the S2 curve for sample J-34 is broader than the other two samples and is associated with a more prominent right-side shoulder plus a smaller, but distinctive, left-side shoulder. The visible "shoulders" in these samples S2-pyrogram curves (Fig. 2.4) is an indication that some traces of reaction products and residues (thermogenic and/or biogenic), as well as kerogens, are present in these samples. Table 2.2 provides a summary of the reaction-kinetic distributions for the three shale samples. Figure 2.5 plots their positions on an *E-lnA* plot relative to the established trend for thermally immature shales.

It is apparent from Table 2.2 and Fig. 2.5 that that the kinetic-reaction distributions for SMR-2B and MJNK-14 have similar weighted average *E* (~247 to 248 kJ/mol) and *lnA* (~66/my) values. On the other hand, sample J-34 displays substantially lower weighted average *E* (~230 kJ/mol) and *lnA* (~62/my) values, and plots slightly further from the established immature shale trend. The weighted average standard deviations for *E* (~65 kJ/mol) and *lnA* (~9/my) of the kinetic distribution for sample J-34 are substantially higher those of the other two shale samples. This is consistent with the broader shape of its S2 curve (Fig. 2.4). The weighted average standard deviation of *E* for sample SMR-2B (~46 kJ/mol) is slightly higher than that of sample MJNK-14 (~43 kJ/mol). Conversely, the weighted average standard deviation of *lnA* for sample SMR-2B (~6/my) is slightly lower than that of sample MJNK-14 (~7/my). The somewhat narrower S2 curve for sample MJNK (Fig. 2.4) is the probable reason for this.

In general, samples of lower thermal maturity are likely to have lower weighted average standard deviations of their reaction-kinetic distributions than more thermally mature samples. This is because there is likely to be a greater range of thermogenic reaction products and residues present in more thermally mature samples

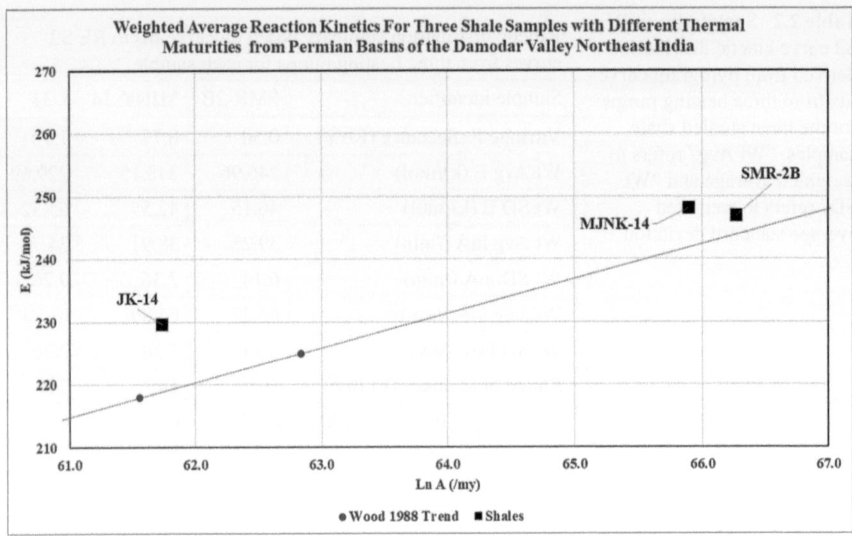

Fig. 2.5 Activation energy (E) versus pre-exponential factor (lnA; expressed in/my units) for the three studied shale samples. The E-A trend which is established for immature organic-rich materials [37] is given for reference

and more complex pore-size distributions in their reacted kerogens. However, in samples affected by biogenic/diagenetic reactions that trend may be disrupted, which may explain the broader fitted E-value distribution of sample SMR-2B compared to sample MJNK-14.

The fitted S2 curves for all three samples are good with low mean-squared errors (mse). The best curve fit is for sample SMR-2B (mse *1000 = 0.3791; Table 2.2). Sample MJNK-14 generates the highest fitting error (mse *1000 = 0.8374; Table 2.2), highlighting that the fitted curve struggled to match the right-shoulder of that sample's S2 pyrogram. Despite its broader and more asymmetrical shape the S2 curve fit for sample J-34 generates a relatively low fitting error (mse *1000 = 0.6804; Table 2.2). Figure 2.6 displays the E values for each of the eleven first-order reactions selected by the optimizer to fit the S2 curves of the three shale samples.

It is apparent from Fig. 2.4 the E-value distributions for samples SMR-2B and MJKN-14 that their reaction-kinetic distributions are unimodal with substantial E peak values of ~235 kJ/mol (SMR-2B) and ~231 kJ/mol (MJKN-14). Those peak reactions account for >40% of the full E distributions selected for the S2 curve fits of those samples, with other reactions contributing <= 10% of the reaction-kinetic distribution. Those peak E values are most likely related to the dominant kerogen type present in those two samples. On the other hand, the E-value distribution for sample J-34 is more complex, as it is bimodal with E peak reactions at ~222 kJ/mol (contributing about 16%) and at ~272 kJ/mol (contributing about 20%).

Assessing the range of the kinetic reactions involved in the fitted S2 curve together with the weighted average standard deviations (E and lnA) and peak E values provide

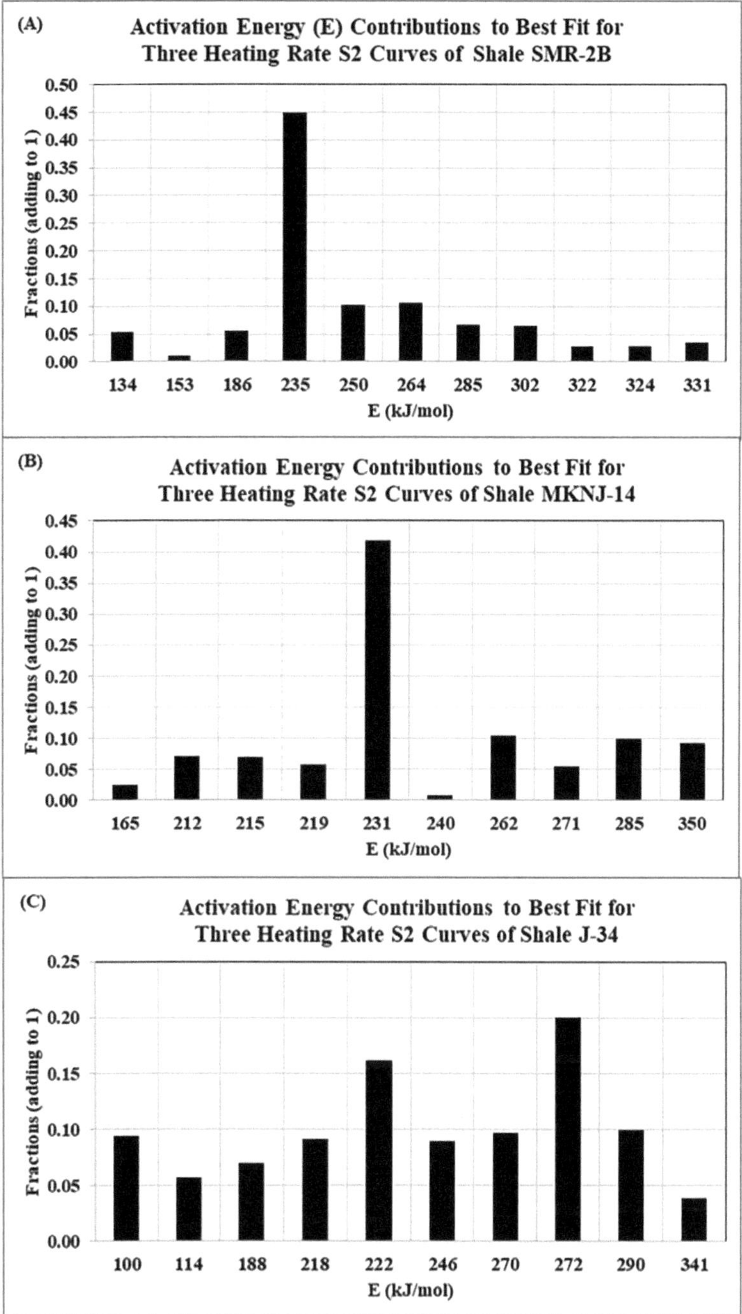

Fig. 2.6 Best fit reaction kinetics distributions calculated from three RE normalized S2 pyrolysis curves generated at heating ramps 25 °C, 15 °C and 5 °C/minute for three studied shale samples **A** SMR-2B (lowest thermal maturity); **B** MKNJ-14; **C** J-34 (highest thermal maturity)

useful insight for characterizing organic-rich shale samples displaying different degrees of thermal maturity. For the three samples studied, the reaction-kinetic distribution of the least mature sample would typically be favored for PSM evaluations to generate burial history *TF* maps and cross-sections for the formation across a basin. This would provide more realistic TF temperatures and depths than applying E-A values often assumed to be representative of generic kerogen types I, II, III, and IV. Such PSM analysis would help to identify the areas where petroleum generation and entrapment would have occurred, and the geological timing of petroleum generation and migration. On the other hand, to identify zones of a shale formation most likely to have high natural gas capacities for exploitation, or those most suitable for injected gas (CH_4 or CO_2) storage, considering samples with S2 pyrograms and reaction-kinetic distributions characteristic of relatively high thermal maturity for the formation is a more appropriate strategy.

2.4 Advances in Geochemical Characterization of Shales Using XRF, XRD, and ICP-MS

A comprehensive view of shale reservoirs goes beyond an understanding of the quality of OM toward a more in-depth study of its mineralogical and elemental composition. Yet although RE pyrolysis continues to be valuable for screening organic content and maturity, these other analytical techniques such as XRD, XRF and ICP-MS have become increasingly vital.

2.4.1 Advances in Application of XRF Analysis

Drawing from advances in technologies used across the earth sciences, XRF analysis has recently become a go-to method for geochemical characterization in shale studies for its non-destructive method, efficient data acquisition, and its ability of analysing both major and trace elements concentrations. Recent advances have broadened its utility past simple elemental profiling, coupling it with machine learning, high-resolution imaging, and multivariate statistics to allow more complex interpretations of shale fabric, depositional environments, and hydrocarbon potential.

One significant recent development is combining XRF with machine learning to allow prediction of OM content. Lawal et al. [16] aimed to advance the characterization of organic-rich shale formations by directly recognising the difficulty of simultaneously analysing their organic (kerogen) and inorganic components. Traditional approaches require stand-alone approaches (mineralogical analysis to determine inorganic content and TOC measurement for OM) that are expensive and labor-intensive. As a more efficient alternative, the researchers suggested using XRF coupled with a machine-learning approach to predict TOC, which XRF alone is unable to measure. The methodology consisted of training an artificial neural network (ANN) to develop a mapping between XRF derived elemental data and corresponding TOC values. The entire dataset was divided into first training (70%), then validation (15%) followed by testing (15%) set for model development and evaluation. Subsequently, a calibration function was extracted from the trained network to allow network driven direct TOC estimation from XRF data. They concluded that ANN model yield an optimum performance for predicting TOC with R^2 equal to 0.974 and MAPE 14.54%. The method is proved to be a cost-effective approach for assessment of organic and inorganic components valuable for both laboratory analysis and field-scale application.

Leveraging the elemental resolution of XRF, chemostratigraphic frameworks can be created to aid sedimentological and resource assessment activities. Zhai et al. [45] showed how chemostratigraphy based on XRF analysis could more efficiently be applied to better characterize sedimentary microfacies and predict shale gas sweet spots within the widely studied Cambrian Niutitang Formation of western Hubei, China. High-resolution elemental data collected using handheld XRF scanning of core samples was able to detect subtle geochemical changes that traditional stratigraphic techniques tend to miss. The quick, highly accurate, and cost-effective elemental profiles created by the XRF technique proved to be the foundation of the multivariate statistical analyses.

Key elemental associations were established through detailed interpretation of data generated using XRF. These comprised Al, Si, S, Ti, V, and Cr for sedimentary facies, Fe, Sr, and Mg for redox conditions, and U, Mo, V, and Ni for organic enrichment. These types made it possible to follow a detailed hierarchy in the classification of the shale lithofacies into four primary facies (Si/Al shale facies, U/Th shale facies, Ca/Mg shale facies and mixed shale facies) that were established and further divided based on fine-scale geochemical patterns. The high-resolution elemental mapping obtained from XRF measurements facilitated the identification of four sedimentary cycles, indicating an evolution from fluvial to marginal marine to marine depositional environments and sea-level rise and fall during the formation's deposition history. Subtle changes in elemental ratios, recorded via high-resolution XRF scans, enabled the authors to identify potential TST shale entries with impressive fidelity across the transgressive systems tract (TST) of the SQ1 sequence.

Optimal sweet spot intervals were forecasted and subsequently validated via drilling activities at the exploratory well EYY1HF which had a 100% net penetration rate through quality shales and resulting gas production performance characterized as outstanding. It's success of this prediction that really highlights just how vital high-resolution XRF data was in connecting geochemical variation to applicable exploration results. This high-detailed, XRF chemostratigraphic dataset allowed for the creation of high-, sequence-, and chemostratigraphic lithofacies/paleogeography maps. These maps provided a compelling visualization of the spatial distribution of sedimentary facies as well as key element concentrations like U and V/(V + Ni).

Mastalerz et al. [23] tested the use of portable x-ray fluorescence (pXRF) elemental analysis to better correlate Pennsylvanian coal-bearing strata in the Indiana part of the Illinois Basin. That study assessed whether the rapid, non-destructive chemical data available from pXRF could help in determining the location of important stratigraphic markers, particularly organic-rich marine shales, and carbonate horizons. The implementation of the pXRF technique played a pivotal role, providing an effective, economical means to efficiently derive major and trace element concentrations in regular intervals of core sections. This catalysed the identification of finely nuanced geochemical signatures associated with specific depositional environments. Uranium and molybdenum elevated concentrations identified at four separate black shale horizons (Sh1–Sh4) which represented important localized anoxic depositional events and acted as useful regional markers. Changes in calcium and magnesium concentrations permitted identification of limestone and dolostone beds, although the survey admitted that distinguishing among distinct carbonate horizons based on elemental data alone was still difficult because of lateral variability. Petrographic analysis was also employed as complementary tool to determine the OM type and OM maturity which worked as additional information for differentiating between the shale horizons.

Similarly, another study observed that important intervals of quartz enrichment and carbonate enrichment zones for hydraulic fracturing potential could be identified using µXRF derived elemental ratios, e.g., Si/Ti, Ca/Ti, Mn/Ti [7]. These high-resolution µXRF profiles in conjunction with PCA analysis allowed for the successful identification of subtle sedimentary variations and depositional features including fining- and coarsening-upward successions, sharp basal contacts, and variations associated with storm events. These features at lower sampling resolution would have been masked under lower sampling resolutions associated with traditional methodologies.

Micro-XRF as a measurement and visualization tool can connect micro- and macro-scale shale heterogeneity characterization [8]. That study introduced a new method to quantitatively describe the meso-scale heterogeneities of the shales. The workflow focused on micro-XRF imaging in order to conduct chemo-sedimentary facies analysis and outline two-dimensional representative elementary areas (REAs). This was efficient method for subsampling and connected micro-scale observations from advanced techniques such as SEM with wireline log data (macro-scale), facilitating a multi-scale characterization framework.

2.4.2 XRD-XRF Integration for Shale Characterization

Integration of XRD and XRF techniques provide the most robust complimentary framework to improve mineral quantification, compositional modelling and facies differentiation, particularly when augmented by machine learning. Hupp and Donovan [13] quantified mineral identification derived from XRD with mineralogically relevant elemental concentrations derived from XRF to further enhance the reliability of the resulting mineralogical assessments. XRD bulk analysis indicated nine primary mineral phases, and XRF yielded exact oxide concentrations, which were divided between the nine mineral phases according to known stoichiometries.

This synergy-based approach enabled mineral abundances to be calculated that were less susceptible to biases often associated with traditional reference-intensity-ratio (RIR) interpretations, and in particular, improvements to the over-abundance of clay minerals and the under-abundance of quartz. Instead of using a systematic mass-balance approach to distribute elemental concentrations across the mineral phases, that study was able to detect subtle compositional differences masked by traditional methods. In the end, this technique provided an internal quality control check through the brilliant match in between predicted and measured Al_2O_5 concentrations that can be seen in the above plot, providing additional validation of the robustness of this approach. Cluster analysis of the XRD-XRF mineralogy data clustered into four distinct mineralogical facies, reflecting differences in depositional environment and primary controls on OM preservation within the shale. Kim et al. [15] observed that combination of XRF-derived elemental information with XRD-based crystallographic identification overcomes the limits of each technique, permitting accurate mineral phase mapping while substantially decreasing the need for tedious diffraction analyses. An ANN model trained on coupled μXRF-μXRD datasets collected from natural Eagle Ford (U.S.A) shale samples and a synthetic mineral mixture achieved classification accuracies above 97%.

Recent advances in field emission scanning electron microscopy (FE-SEM) for imaging shales coupled with deep-learning (deep-learning, and specifically KiU-Net configurations of convolutional neural networks) has resulted in a cost-effective digital core methodology for micro-scale shale characterization [44]. So far this process has been limited to mineral identification to the pinpoint level and identification of micro-distribution mineral-assemblages, using a quantitative evaluation of minerals by FESEM (QEMSCAN; high-resolution mineral analysis) data sets. Hybrid deep learning models such as deep belief network (DBN) and long short-term memory (LSTM) have also been successfully applied to predict TOC and vitrinite reflectance data from well logs in complex lacustrine oil shale formations [20]. Potential exists to combine such deep-learning shale characterization method with μXRF-μXRD datasets to provide more geochemically focused digital cores and micro-scale characterizations.

2.4.3 Advances in Application of ICP-MS Analysis

Inductively coupled plasma mass spectrometry (ICP-MS) has recently been useful in analysing the interactions between hydraulic fracturing fluids and shale matrices. Dje et al. [6] using ICP-MS studied shale interaction with hydraulic fracturing fluids. The high sensitivity of ICP-MS greatly aided the detection of subtle elemental variations that mirrored not only mineral dissolution, but additionally precipitation of secondary, complexed and amorphous minerals. Ca and Si concentrations rose over time due to dissolution of carbonate and silicate minerals. Al and Fe showed negative or relatively constant trends possibly due to uptake of secondary phases. The technique enabled reconstruction of geochemical pathways through the shale matrix.

Beyond tracking mineralogical transformations, that study used ICP-MS to assess environmental risks posed by flowback water. While Caney shale effluents had overall lower concentrations of As and Pb compared to Marcellus shale effluents, high concentrations of B and Se were found (both of which increase irrefutable concerns over water quality). The accuracy and precision quantification possible with ICP-MS was essential for evaluation of effluent concentrations with respect to regulatory limits. The implementation of ICP-MS allowed for detailed investigation of shale reactivity and effluent chemistry, providing a wealth of information on the effects of hydraulic fracturing fluids into reservoir integrity and ecological safety. The technique allowed for an unprecedented view of subsurface geochemical processes at work and possible implications at the surface.

Awejori et al. [2] assessed the geochemical evolution of the Caney Shale under conditions simulating post-hydraulic fracturing fluid-rock interactions, with ramifications for both hydrocarbon recovery and carbon storage. A key analytical tool used in the study was ICP-MS, which allowed for precise quantification of elemental concentrations in experimental effluents. Using high-resolution ICP-MS, that work was able to track the release and uptake of key elements Si, Al, Fe, Ca, Sr, Ba, Mg, and Na in time, allowing for complex mineral dissolution–precipitation dynamics to be reconstructed over 7- and 30-day reaction periods. Based on the ICP-MS results, the study determined that early pyrite oxidation by oxygenated, fracturing fluids created localized acid conditions that later triggered feldspar and carbonate minerals dissolution.

Elemental signatures recorded via ICP-MS analyses in that research suggested important mobilization of cations and anions related to mineral weathering, through secondary mineral precipitation reactions harmful to reservoir permeability. In particular, the targeted concentrations of Fe, Sr, and Ba distinguished the precipitation of Fe(III) hydroxides, celestite ($SrSO_4$), and Barite ($BaSO_4$) scales. These scales are naturally effective at chiselling away at fracture conductivity and wonderfully effective at hastening permeability decline, effects that would otherwise be challenging to identify or measure without the resolution powers of ICP-MS.

This fine-scale elemental data harnessed from ICP-MS was able to better illustrate mineralogical changes observed through XRD, creating a dynamic, detailed perspective on the advancing rock-fluid system. Complementary geochemical modelling, supported by ICP-MS measurements, further confirmed dissolution–precipitation pathways seen in experiments, highlighting the sensitivity of shale reservoirs to variation in fluid composition and redox state. That earlier study showed that ICP-MS is not just good for diagnosing mechanisms of formation damage but useful for guiding optimization of fracture-fluid compositions to avoid adverse geochemical reactions.

Jew and Brownlow [14] combined ICP-MS and inductively coupled plasma optical emission spectroscopy (ICP-OES) to examine trace and major element concentrations in model solutions over time. ICP-MS identified fundamental geochemical processes such as leaching of Sr when reacted with freshwater and precipitation of celestite ($SrSO_4$) when clean brine was applied. Multi-dimensional chemical monitoring made possible by ICP-MS revealed similar dynamic behavior of iron, Fe cycling quickly from a dissolved state to a precipitated state, resulting in the formation of reactive Fe(III) oxides without detectable changes in bulk mineralogy. ICP-MS data were critical in determining how additive decomposition (e.g., persulfate breakers) and oxidation of native minerals contributed to both sulfate scaling and secondary mineral formation, processes that bulk analyses may miss.

In summary, the implementation of ICP-MS into the experimental workflow greatly enriches our ability to trace nuanced but impactful geochemical changes within shale matrices. In addition to being a tool for determining chemical signature, high-sensitivity IPC-MS elemental analysis can be applied to characterize complex inorganic reactions that are prevalent in unconventional reservoirs.

2.4.4 Application of Geochemical Proxies in Predicting Shale Geomechanical Properties

Characterizing the geomechanical properties of shale formations has been a tremendous task given the characteristically fissile and laminated nature of shales [28, 29]. These traits create challenges for the retrieval of wide, continuous core samples needed for traditional mechanical testing. Thus, characterization of strength and deformation parameters for specific layers of a shale formation that are fractured possesses several challenges. Therefore, indirect methods such as use of geochemical proxies for prediction of shale mechanical behavior should be considered. Recent work developing geomechanical properties of shales from geochemical properties demonstrated the feasibility of this approach. Li et al. [17] conducted a series of laboratory experiments to derive geochemical indicators, which are linked with the depositional environment of the shale and can be used to infer its tensile strength. Centering on shales from the Banjiuguan Formation in the Micangshan Mountain area (China), that study used a suite of geochemical analyses (XRF, ICP-MS, TOC) and mechanical testing (Brazilian tensile strength, acoustic emission) to explore the manner in which paleoenvironmental processes controlled shale strength. In the most significant findings, tensile strength valued from 14.03 to 25.62 MPa, with a mean value of 20.84 MPa in the studied samples. TOC was only weakly negatively correlated with tensile strength and was not the main controlling variable. Instead, it was mineral compositions and sedimentary conditions that were of greater deciding factors. In particular, high paleoproductivity and biogenic silica input, characteristic of a marine continental margin environment, positively influenced quartz content, increasing the shales brittleness and strength. Redox paleoclimate conditions were more likely to lower tensile strength. Weathering under warm and humid conditions made it easier for carbonate minerals to dissolve, creating a greater porosity but compromised structural integrity of shales. Anoxic conditions, while conducive to OM preservation, restricted formation of strengthening minerals like pyrite, rendering the rock more vulnerable to failure. Palaeosalinity was a large biological controlling factor as well. Advanced salinity levels at deposition likely encouraged chemical conditions which led to the development of a tighter clay-quartz framework. This structural compaction corresponded to the increased mechanical strength, establishing palaeosalinity as a compaction-enforcing mechano agent.

Similarly, Sethi et al. [28] assessed geochemical proxies to determine the mechanical behavior of Rajmahal shales. BTS tests showed tensile strengths between 0.93 and 4.12 MPa and the brittleness index (BI) from 0.71 to 3.40. TOC showed a clear threshold effect: samples with TOC >15 wt% displayed a significant negative correlation with tensile strength, indicating the weakening role of excess OM in the shale matrix (Fig. 2.7A and B). BI exhibited very weak correlation with TOC content (Fig. 2.7C and D). Thermal maturity, determined via T_{max}, had little effect on the tensile strength or BI for the samples studied, since all samples were in a fairly close maturity range (Fig. 2.8).

Mineralogical composition made a greater impact. Clay mineral (CM) content was positively correlated with tensile strength and BI (Fig. 2.9A and C), while quartz was negatively correlated with these two properties (Fig. 2.9B and D). This is opposite what many past studies have found, that quartz increases the brittleness, hinting that the type and source of quartz could be an important consideration.

Indeed, the source of silica turned out to be a determining factor. Samples rich in biogenic silica, identified from SiO_2/Al_2O_5 ratios >3 and backed by mineralogy, are usually less tensile and more brittle than samples composed of lithogenic silica

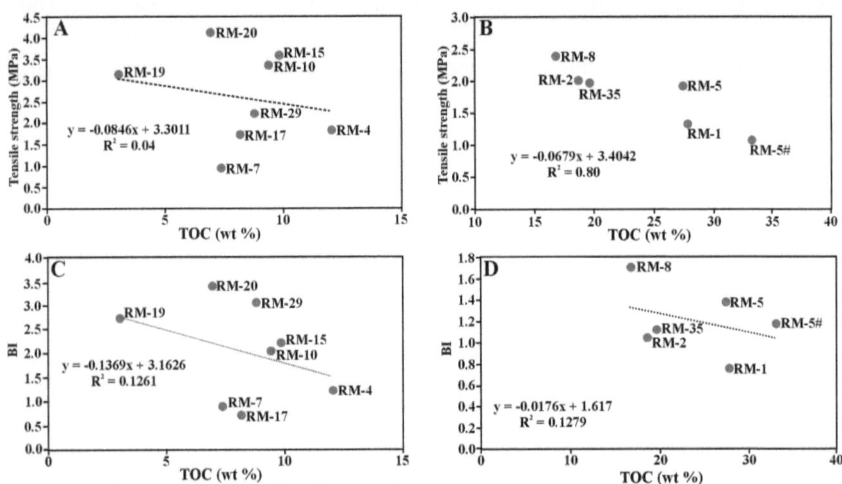

Fig. 2.7 **A** and **B** Tensile strength & **C** and **D** Brittleness index versus TOC of RB shale samples. (Reprinted from Sethi, C., Hazra, B., Ostadhassan, M., Motra, H.B., Dutta, A., Pandey, J.K. and Kumar, S., "Depositional environmental controls on mechanical stratigraphy of Barakar Shales in Rajmahal Basin, India," *International Journal of Coal Geology*, Vol. 285, p. 104477, Copyright (2024), under the terms of the Creative Commons Attribution License (CC BY 4.0). https://creati vecommons.org/licenses/by/4.0/)

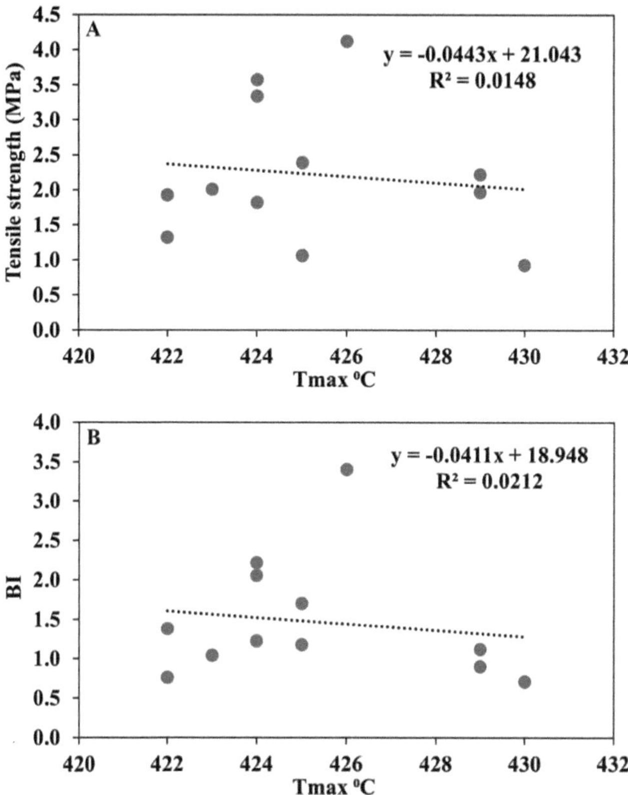

Fig. 2.8 **A** Tensile strength and **B** Brittleness index versus T_{max} of shales. (Reprinted from Sethi, C., Hazra, B., Ostadhassan, M., Motra, H.B., Dutta, A., Pandey, J.K. and Kumar, S., "Depositional environmental controls on mechanical stratigraphy of Barakar Shales in Rajmahal Basin, India," *International Journal of Coal Geology*, Vol. 285, p. 104477, Copyright (2024), under the terms of the Creative Commons Attribution License (CC BY 4.0). https://creativecommons.org/licenses/by/4.0/)

(Fig. 2.10A and B). This trend was opposite to what would be expected classically, as biogenic quartz generally increases brittleness because of its microcrystalline structure. The predominance of CM in the samples seemed to eclipse quartz's contributions to mechanical behavior.

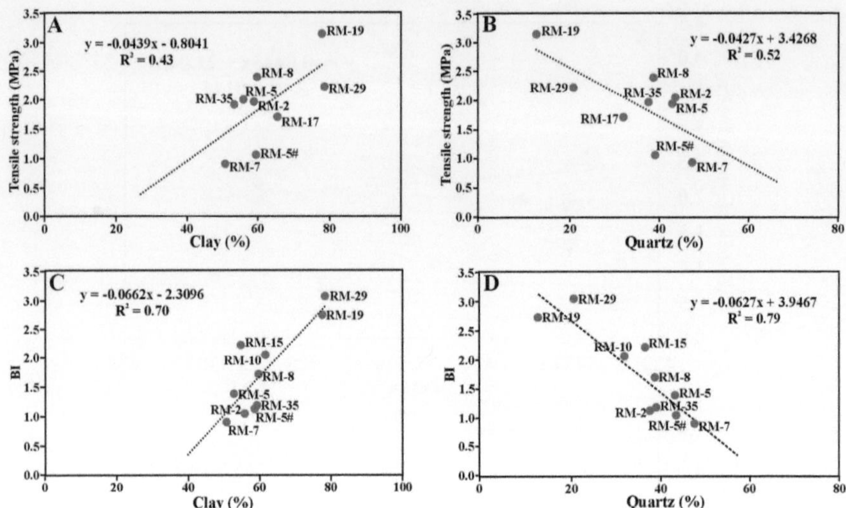

Fig. 2.9 Correlations between geomechanical properties and mineralogical composition of RB shales. (Reprinted from Sethi, C., Hazra, B., Ostadhassan, M., Motra, H.B., Dutta, A., Pandey, J.K. and Kumar, S., "Depositional environmental controls on mechanical stratigraphy of Barakar Shales in Rajmahal Basin, India," *International Journal of Coal Geology*, Vol. 285, p. 104477, Copyright (2024), under the terms of the Creative Commons Attribution License (CC BY 4.0). https://creati vecommons.org/licenses/by/4.0/)

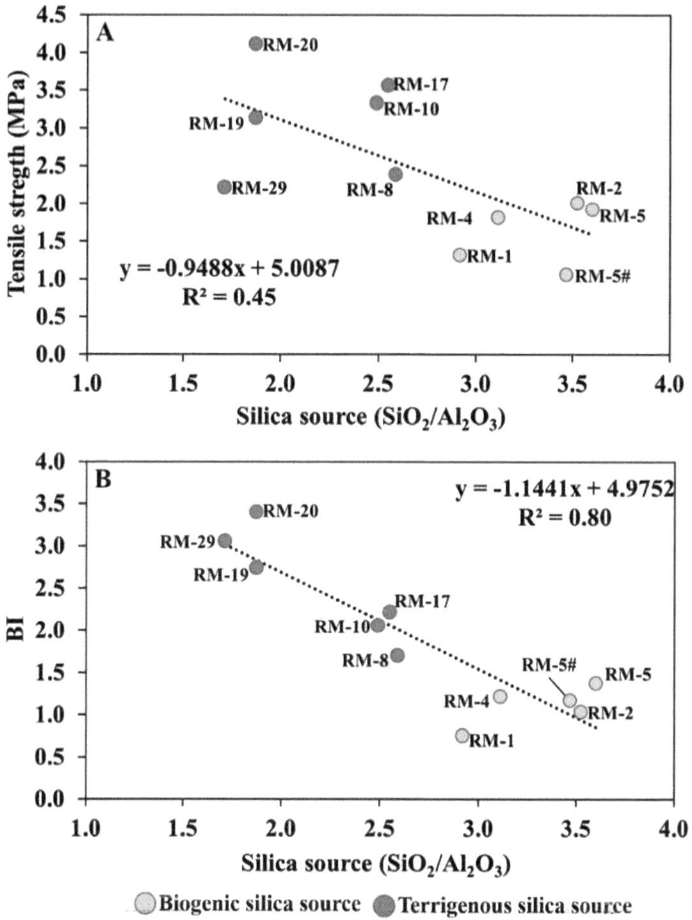

Fig. 2.10 Relationship between silica source and shale properties. **A** Tensile strength and **B** brittleness index (BI) of the shale samples plotted against the silica source, represented by the SiO_2/Al_2O_3 ratio. (Reprinted from Sethi, C., Hazra, B., Ostadhassan, M., Motra, H.B., Dutta, A., Pandey, J.K. and Kumar, S., "Depositional environmental controls on mechanical stratigraphy of Barakar Shales in Rajmahal Basin, India," *International Journal of Coal Geology*, Vol. 285, p. 104477, Copyright (2024), under the terms of the Creative Commons Attribution License (CC BY 4.0). https://creati vecommons.org/licenses/by/4.0/)

References

1. Arrhenius S (1889) Über die Reaktionsgeschwindigkeit bei der Inversion von Rohrzucker durch Säuren. Z Phys Chem 4:226–248
2. Awejori GA, Dong W, Doughty C, Spycher N, Radonjic M (2024) Mineralogy and reactive fluid chemistry evolution of hydraulically fractured Caney shale of Southern Oklahoma. Gas Sci Eng 131:205458

3. Badejo SA, Fraser AJ, Neumaier M, Muxworthy AR, Perkins JR (2021) 3D petroleum systems modelling as an exploration tool in mature basins: a study from the Central North Sea UK. Mar Pet Geol 133:105271. https://doi.org/10.1016/j.marpetgeo.2021.105271

4. Behar F, Beaumont VDEB, Penteado HDB (2001) Rock-Eval 6 technology: performances and developments. Oil Gas Sci Technol 56(2):111–134

5. Chen Z, Liu X, Guo Q, Jiang C, Mort A (2017) Inversion of source rock hydrocarbon generation kinetics from Rock-Eval data. Fuel 194:91–101. https://doi.org/10.1016/j.fuel.2016.12.052

6. Dje LB, Awejori GA, Radonjic M (2024) Comparison of geochemical reactivity of marcellus and caney shale based on effluent analysis. In: ARMA US rock mechanics/geomechanics symposium, p D041S053R004. ARMA

7. Gabriel JJ, Reinhardt EG, Chang X, Bhattacharya JP (2022) Application of μXRF analysis on the Upper Cretaceous Mancos Shale: a comparison with ICP-OES/MS. Mar Pet Geol 140:105662

8. Guo Z, Liu K, Yu L, Chen J, Wu Y, Chen L, Liang C (2025) Characterizing multi-scale heterogeneities in shales: an innovative workflow and application to Paleogene lacustrine shales in the Subei Basin, eastern China. Mar Pet Geol 107398

9. Hazra B, Chandra D, Vishal V (2024) Unconventional hydrocarbon reservoirs: coal and shale. Springer

10. Hazra B, Katz BJ, Singh DP, Singh PK (2022) Impact of siderite on Rock-Eval S3 and oxygen index. Mar Pet Geol 143:105804

11. Hazra B, Singh DP, Chakraborty P, Singh PK, Sahu SG, Adak AK (2021) Using rock-eval S4Tpeak as thermal maturity proxy for shales. Mar Pet Geol 127:104977

12. Hazra B, Wood DA, Mani D, Singh PK, Singh AK (2019) Source-rock geochemistry: organic content, type, and maturity. In: Evaluation of shale source rocks and reservoirs, pp 7–17 and 85–105

13. Hupp BN, Donovan JJ (2018) Quantitative mineralogy for facies definition in the Marcellus Shale (Appalachian Basin, USA) using XRD-XRF integration. Sed Geol 371:16–31

14. Jew AD, Brownlow JW (2025) Inorganic alterations in unconventional shale reservoirs: importance of additive and base fluid chemistry. Energy & Fuels

15. Kim JJ, Ling FT, Plattenberger DA, Clarens AF, Lanzirotti A, Newville M, Peters CA (2021) SMART mineral mapping: synchrotron-based machine learning approach for 2D characterization with coupled micro XRF-XRD. Comput Geosci 156:104898

16. Lawal LO, Mahmoud M, Alade OS, Abdulraheem A (2019) Total organic carbon characterization using neural-network analysis of XRF data. Petrophysics 60(04):480–493

17. Li H, Li D, He Q, Sun Q, Zhao X (2024) Controlling mechanism of shale palaeoenvironment on its tensile strength: a case study of Banjiuguan Formation in Micangshan Mountain. Fuel 355:129505

18. Li Q, Chen F, Wu S, Zhang L, Wang Y, Xu S (2022) A simple and effective evaluation method for lacustrine shale oil based on mass balance calculation of Rock-Eval data. Appl Geochem 140:105287

19. Liao L, Wang Y, Chen C, Pan Y (2022) Application of the grain-based Rock-Eval pyrolysis method to evaluate hydrocarbon generation, expulsion, and retention of lacustrine shale. Front Earth Sci 10:921806

20. Liu B, Ma Y, Qamar Y, Wood DA, Sun M, Gao S, Bai Y (2025) Characterization of lacustrine shale oil reservoirs based on a hybrid deep learning model: a data-driven approach to predict lithofacies, vitrinite reflectance, and TOC. Mar Pet Geol 174:107309. https://doi.org/10.1016/j.marpetgeo.2025.107309

21. Ma Y, Li S (2018) The mechanism and kinetics of oil shale pyrolysis in the presence of water. Carbon Resour Convers 1(2):160–164. https://doi.org/10.1016/j.crcon.2018.04.003

22. Mastalerz M, Drobniak A, Stankiewicz A (2018) Origin, properties, and implications of solid bitumen in source-rock reservoirs: a review. Int J Coal Geol 195:14–36. https://doi.org/10.1016/j.coal.2018.05.013

23. Mastalerz M, Drobniak A, Ames P, McLaughlin PI (2019) Application of pXRF elemental analysis and organic petrography in correlation of Pennsylvanian strata: an example from the Indiana part of the Illinois Basin, USA. Int J Coal Geol 216:103342

24. Peters KE, Burnham AK, Walters CC (2015) Petroleum generation kinetics: single versus multiple heating-ramp open-system pyrolysis. AAPG Bull 99(4):591–616

25. Peters KE, Burnham AK, Walters CC, Schenk O (2018) Guidelines for kinetic input to basin and petroleum system models. Mar Pet Geol 92:979–986. https://doi.org/10.1016/j.marpetgeo.2017.11.024

26. Schneider F, Dubille M, Montadert L (2016) Modeling of microbial gas generation: application to the eastern Mediterranean "Biogenic Play." Geol Acta 14(4):403–417. https://doi.org/10.1344/GeologicaActa2016.14.4.5

27. Sethi C, Hazra B (2022) Source rock properties of Permian shales from Rajmahal basin, India. Arab J Geosci 15(13):1219

28. Sethi C, Hazra B, Ostadhassan M, Motra HB, Dutta A, Pandey JK, Kumar S (2024) Depositional environmental controls on mechanical stratigraphy of Barakar shales in Rajmahal basin, India. Int J Coal Geol 285:104477

29. Sethi C, Hazra B, Wood DA, Singh AK (2023) Experimental protocols to determine reliable organic geochemistry and geomechanical screening criteria for shales. J Earth Syst Sci 132(2):45. (Custrine shale. Front Earth Sci 10:921806)

30. Shabani F, Amini A, Tavakoli V, Chehrazi A, Gong C (2023) 3D Basin and petroleum system modelling of the early Cretaceous play in the NW Persian Gulf. Geoenergy Sci Eng 226:211768. https://doi.org/10.1016/j.geoen.2023.211768

31. Singh DP, Wood DA, Hazra B, Singh PK (2021) Pore properties in organic-rich shales derived using multiple fractal determination models applied to two Indian Permian Basins. Energy Fuels 35(18):14618–14633. https://doi.org/10.1021/acs.energyfuels.1c02077

32. Singh DP, Wood DA, Singh V, Hazra B, Singh PK (2022) Impact of particle crush-size and weight on Rock-Eval S2, S4, and kinetics of shales. J Earth Sci 33(2):513–524

33. Tissot BP, Welte DH (1984) Petroleum formation and occurrence; a new approach to oil and gas exploration (second edition), p 702. Springer-Verlag, Berlin, New York. https://doi.org/10.1007/978-3-642-87813-8

34. Ungerer P, Pelet R (1987) Extrapolation of the kinetics of oil and gas formationfrom laboratory experiments to sedimentary basins. Nature 327:42–54. https://doi.org/10.1038/327052a0

35. Wood DA (2018) Thermal maturity and burial history modelling of shale is enhanced by use of Arrhenius time-temperature index and memetic optimizer. Petroleum 4:25–42

36. Wood DA (2018) Kerogen conversion and thermal maturity modelling of petroleum generation: integrated analysis applying relevant kerogen kinetics. Mar Pet Geol 89:313–329. https://doi.org/10.1016/j.marpetgeo.2017.10.003

37. Wood DA (1988) Relationships between thermal maturity indices calculated using Arrhenius equation and Lopatin method: implications for petroleum exploration. AAPG Bull 72(2):115–135. https://doi.org/10.1306/703C8263-1707-11D7-8645000102C1865D

38. Wood DA (2017) Re-establishing the merits of thermal maturity and petroleum generation multi-dimensional modelling with an Arrhenius equation using a single activation energy. J Earth Sci 28(5):804–834. https://doi.org/10.1007/s12583-017-0735-7

39. Wood DA (2019) Establishing credible reaction-kinetics distributions to fit and explain multi-heating rate S2 pyrolysis peaks of kerogens and shales. Adv Geo-Energy Res 3(1):1–28. https://doi.org/10.26804/ager.2019.01.01

40. Wood DA (2022) Shale kerogen kinetics from multi-heating rate pyrolysis modelling with geological time-scale perspectives for petroleum generation. In: Wood DA, Cai J (eds) Sustainable geoscience for natural gas sub-surface systems, pp. 159–195. Elsevier. https://doi.org/10.1016/B978-0-323-85465-8.00001-7. (Chapter 6)

41. Wood DA (2024) Kerogen kinetic distributions and simulations provide insights into petroleum transformation fraction (TF) profiles of organic-rich shales. J Earth Sci 35:747–757. https://doi.org/10.1007/s12583-024-1981-0

42. Wood DA, Hazra B (2018) Pyrolysis S2-peak characteristics of Raniganj shales (India) reflect complex combinations of kerogen kinetics and other processes related to different levels of thermal maturity. Adv Geo-Energy Res 2(4):343–368. https://doi.org/10.26804/ager.2018.04.01

43. Wood DA, Hazra B (2017) Characterization of organic-rich shales for petroleum exploration & exploitation: a review- part 2: geochemistry, thermal maturity, isotopes and biomarkers. J Earth Sci 28(5):758–778
44. Yasin Q, Liu B, Sun M, Sohail GM, Ismail A, Wood DA, Chen Z, Wang Z (2025) Digital core modeling for multimineral segmentation of lacustrine shale oil using FE-SEM and KiU-Net. Fuel 398:135474. https://doi.org/10.1016/j.fuel.2025.135474
45. Zhai G, Li J, Jiao Y, Wang Y, Liu G, Xu Q, Wang C, Chen R, Guo X (2019) Applications of chemostratigraphy in a characterization of shale gas Sedimentary Microfacies and predictions of sweet spots—taking the Cambrian black shales in Western Hubei as an example. Mar Pet Geol 109:547–560

Chapter 3
Combining Multi-scale Imaging and Spectroscopy for Shale Characterization

Abstract Several different microscopy techniques optical, Optical-electron correlative microscopy (OECM), Raman and infrared (IR) spectroscopy, atomic force microscopy (AFM), X-ray computed tomography (X-ray CT), and focused ion beam-scanning electron microscopy (FIB-SEM) when used in combination reveal important and complementary details of shales at multiple scales and resolutions. OECM is effective at distinguishing variations in porosity distributions of shales and how they change as thermal maturity progresses, and relate these changes with specific organic matter type. AFM can provide high-resolution images at the nanoscale, providing insights in how shale pore structure, mineral distributions and micro-mechanical properties varies as thermal maturity progresses. AFM-IR is useful for assessing nano-scale geochemical and mechanical properties at much improved resolutions compared to IR used in isolation. X-ray CT images can discern pore-system features and assess shale anisotropy/heterogeneity at multiple scales. Deep-learning techniques, particularly various configuration of convolutional neural networks (CNN), are effective for CT image segmentation to distinguish features of interest and assess their distributions. Various super-resolution techniques are able to enhance the resolutions of low-resolution recorded CT images to discern very-small features. Applied to FIB-SEM images of various shale formations, the KiU-Net multi-branched configuration of CNN has shown its capability to resolve nanopores and micro-fractures in the size range 10–100 nm. These techniques offer the potential to substantially expand the level of detail extracted at multiple scales from shale images providing more effective formation characterization.

Keywords Optical-electron correlative microscopy (OECM) · Infrared (IR) spectroscopy · Atomic force microscopy (AFM) · X-ray computed tomography (X-ray CT) · Focused ion beam—scanning electron microscopy (FIB-SEM) · Convolutional neural networks (CNN) · KiU-Net CNN · Micro-mechanical nano-scale properties · Chemical gradients in organic matter

C. Sethi et al., *Recent Advances in Shale Characterization Based on Laboratory Geochemistry and Geomechanics Techniques*, SpringerBriefs in Petroleum Geoscience & Engineering, https://doi.org/10.1007/978-3-032-03961-3_3

3.1   Introduction

Application of high-resolution imaging and spectroscopy techniques are necessary for the characterization of the complex microstructure of shales [7]. Traditionally, optical microscopy was used to study organic matter (OM) in shales, however, recent advancements in correlative microscopy, Raman and infrared spectroscopy, atomic force microscopy (AFM), X-ray computed tomography (X-ray CT), and focused ion beam—scanning electron microscopy (FIB-SEM) have enabled researchers to analyse shales from multi-scale perspective.

3.2   Optical Microscopy and Organic Matter Identification

Optical microscopy technique enables classification of macerals (vitrinite, inertinite, and liptinite) and evaluation of OM composition and maturity [12, 15, 23, 34]. Knowledge about the distribution and abundance of OM in shales is critical since varying types of OM has differing potential for hydrocarbon generation [13, 14]. Oil-prone shale formations are generally liptinite rich, whereas gas producing formations are vitrinite rich. This geochemical composition analysis paired with thermal maturity assessment allows for a more thorough and accurate evaluation of a shale reservoir's potential.

Some disadvantages of optical microscopy include relatively low resolution, limited by the diffraction limit, prevents them from being able to observe nanoscale features that are key to understanding the chemical composition and microstructural arrangement of dispersed organic matter (DOM) within the mineral matrix. Usually, the objective lenses have a magnification of about $4\times$ to $100\times$, and the eyepieces add more magnification, usually $10\times$. This additional combination reaches total magnifications up to about $1000\times$ [34]. This shortcoming requires the combination of advanced high-resolution imaging techniques, including but not limited to scanning electron microscopy (SEM) and AFM, in order to provide more in-depth analysis of shale's complicated microstructure.

3.3   Optical-Electron Correlative Microscopy

Optical-electron correlative microscopy (OECM) is an emerging technique that combines optical microscopy and SEM to determine the morphology, distribution and spatial relations of OM and mineral phases in shales [35, 42]. The approach has gained growing popularity in recent years especially in the characterization of unconventional petroleum systems, as scientists and engineers have tried to break the bounds of independent imaging techniques. As discussed earlier, optical microscopy is limited by its resolution. Although, SEM can get to higher resolution for characterization of shale microstructure it cannot distinguish between different OM types as they appear dark under SEM due to low atomic number [5, 25]. The introduction of OECM has given us a tremendous breakthrough in addressing the above limitations by allowing direct linking of petrographic observations with nanoscale structural characterization. This innovative technique combines OM type identification under optical microscopy, and then accurate repositioning of the same particles under SEM for more in-depth analysis [12, 35]. The capacity to scrutinize the same regions of interest (ROIs) on both modalities permits organic components to be accurately classified and further analysed at significantly higher resolutions. These advances have generated more reproducible interpretations of OM-hosted structural porosity, OM-mineral reactions, and thermally induced structural alteration due to maturity.

3.3.1 Correlative Technique

The OECM method consists of a set of systematic, iterative steps that are specifically aimed at facilitating optical-electron microscopy fusion. The initial step in the imaging process is sample preparation, during which shale samples are embedded in resin and polished to a sufficient flatness to allow for both optical and SEM imaging. The samples for petrographic analysis were prepared following the established standards and then mounted onto a Shuttle & Find (S&F) sample holder (Fig. 3.1). The holder is designed in a way that it enables smooth transfer between the microscopes. Samples, obtained this way, are first analysed under optical microscopy in reflected white light and fluorescence illumination. This makes it possible to identify maceral types by their optical properties [16, 17].

Once the OM is identified and marked, the next step is to annotate regions of interest (ROIs) with high precision using digital analysis software. These ROIs are rigorously defined to guarantee their exact re-mapping under SEM. This sample is then moved into SEM chamber. During this time, a very thin (~10–15 nm thick) conductive coating (usually of gold or carbon) is applied to the sample to avoid charging effects and enhance the quality of the imaging. The Shuttle & Find system enables automatic relocation of the marked ROIs, ensuring that the exact same OM particles are analysed under SEM. Secondary electron (SE) and backscattered electron (BSE) scanning modes of SEM are used to study the microstructural properties of OM and mineral phases. Energy-Dispersive X-ray spectroscopy is also applied to determine the elemental composition of the organic and mineral phases.

Sethi et al. [35] used OECM technique and established new protocols to study the microstructural properties of OM and mineral phases in shales of contrasting thermal maturity. They found that using BSE scanning mode at 15 kV (Fig. 3.2C) and SE2 scanning mode at 10 kV (Fig. 3.2F) enables a clear distinction between these two phases. Even at 10 kV under BSE mode (Fig. 3.2E–H), the distinction between organic and mineral matter is not always clearly visible, with contrast further reduced at 5 kV (Fig. 3.2G–H). Likewise, under SE2 mode, both 15 kV (Fig. 3.2D) and 5 kV

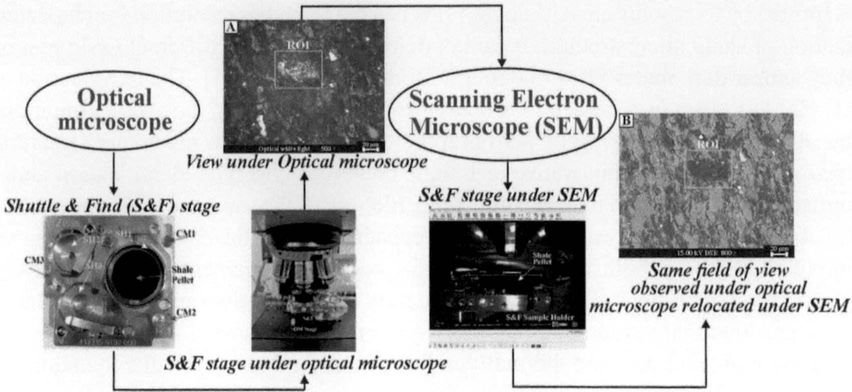

Fig. 3.1 Schematic of correlative microscopy procedure (Adapted from *International Journal of Coal Geology*, Vol. 274, Sethi, C., Mastalerz, M., Hower, J.C., Hazra, B., Singh, A.K. and Vishal, V., "Using optical-electron correlative microscopy for shales of contrasting thermal maturity," p. 104273, Copyright (2023), with permission from Elsevier. License Number: 6030310456162)

(Fig. 3.2H) produce low contrast as well. Thus, it is difficult to identify the phases. Photomicrographs acquired at different working distances (WD) (11, 13, and 15 mm WD) were examined with the conclusion that a working distance of 13 mm yielded the most consistent results when separating DOM from mineral matter in the SEM instrument used for this work (Fig. 3.3).

3.3.2 OECM-Based Multi-scale Analysis of Shale Microstructure

Correlative imaging enabled researchers to correlate the high-resolution microstructural properties of dispersed organic matter in shales with the specific maceral type. This enabled understanding of the impact of thermal maturity on maceral specific porosity evolution and interaction between the organic phase and mineral phase in the shale matrix [35, 42].

Valentine and Hackley [42] used the correlative method to describe OM in North American shale petroleum systems. A key control on the ability of shale reservoirs to both store and transmit hydrocarbons is OM-hosted porosity. As catagenesis continues, OM experiences an increase of structural changes that lead to the development of nanopores, acting as storage space to hydrocarbons. Yet, the degree of organic porosity evolution shows considerable variation between maceral types with thermal maturity. In their study, correlative microscopy allowed for new understanding of how these processes take place, through the ability to precisely relocate

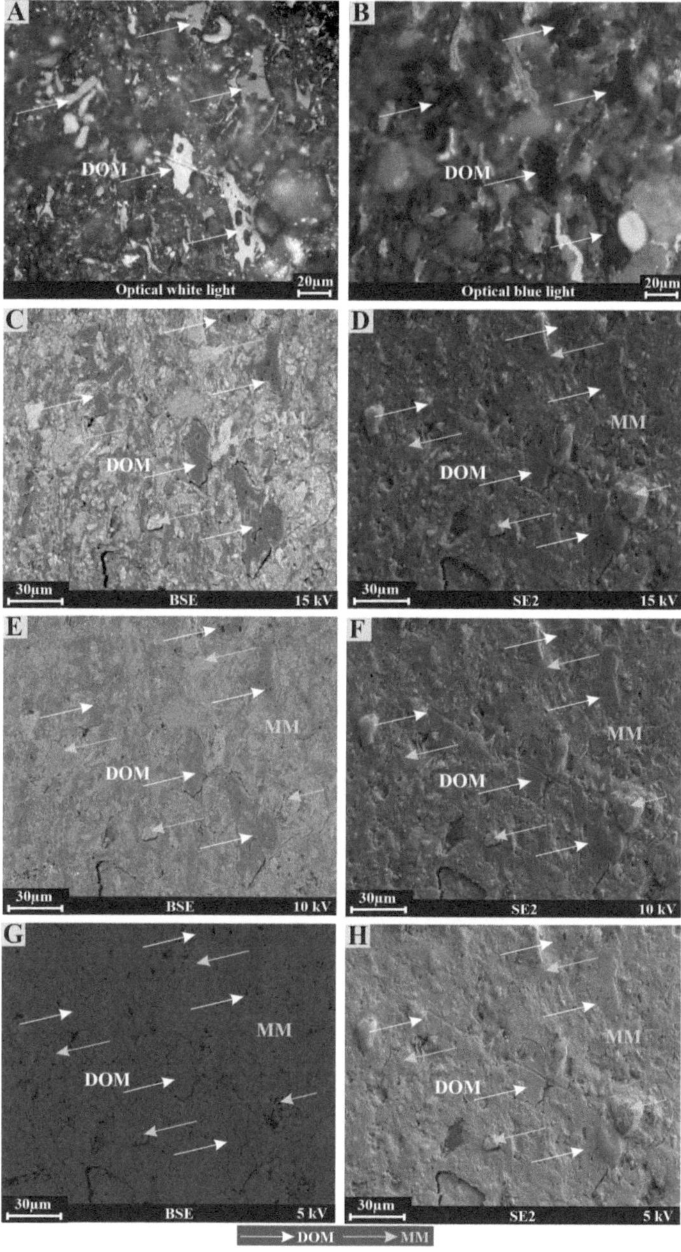

Fig. 3.2 Correlative imaging of the RM shale sample captured under various conditions: **A** bright-field reflected light, **B** fluorescence light, **C** BSE imaging at 15 kV with 665× magnification, **D** SE2 imaging at 15 kV with 665 × magnification, **E** BSE imaging at 10 kV with 665× magnification, **F** SE2 imaging at 10 kV with 665 × magnification, **G** BSE imaging at 5 kV with 665× magnification, and **H** SE2 imaging at 5 kV with 665× magnification. Abbreviations: DOM—Dispersed organic matter; MM—Mineral matter. (Reprinted from *International Journal of Coal Geology*, Vol. 274, Sethi, C., Mastalerz, M., Hower, J.C., Hazra, B., Singh, A.K. and Vishal, V., "Using optical-electron correlative microscopy for shales of contrasting thermal maturity," p. 104273, Copyright (2023), with permission from Elsevier. License Number: 6030310456162)

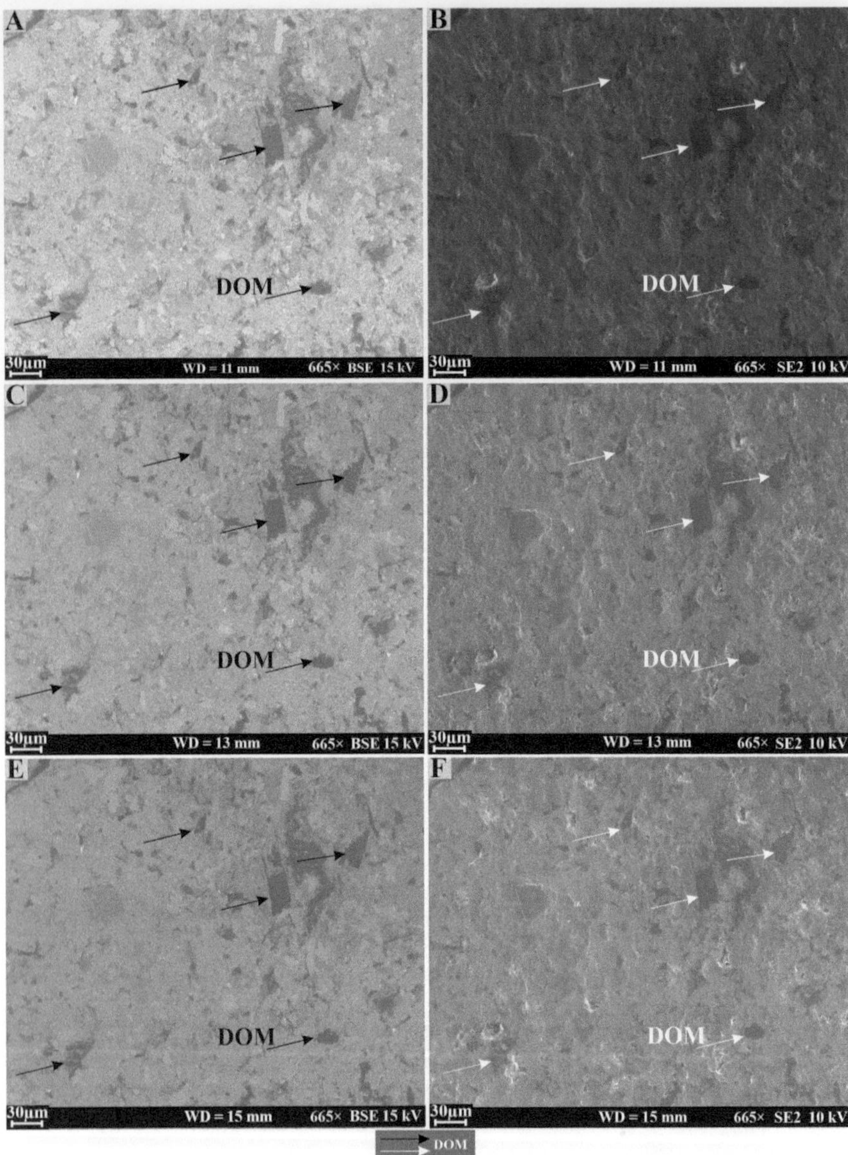

Fig. 3.3 Shale SEM photomicrographs captured at varying WD: 11 mm WD (**A**, **B**), 13 mm WD (**B**, **C**), and 15 mm WD (**C**, **D**). Abbreviation: DOM—Dispersed organic matter. (Reprinted from *International Journal of Coal Geology*, Vol. 274, Sethi, C., Mastalerz, M., Hower, J.C., Hazra, B., Singh, A.K. and Vishal, V., "Using optical-electron correlative microscopy for shales of contrasting thermal maturity," p. 104273, Copyright (2023), with permission from Elsevier. License Number: 6030310456162)

OM visualized under optical microscopy for subsequent high-resolution SEM characterization. It was observed that the porosity development in OM was of varying nature for different shale formations.

In the Woodford Shale, they found that bitumen was well distributed across the shale cores and were only isolated to low amounts of inertinite. Optical microscopy was important for differentiating these materials by optical reflectance measurement. Optical imaging in conjunction with SEM imaging was able to confirm that these nanopores in fact were primarily found within the solid bitumen. Without the correlative approach, based on SEM observations alone, confusion could have arisen between inertinite and bitumen, leading to erroneous conclusions regarding porosity distribution. Haynesville Formation hosted solid bitumen-filled voids to terrestrial kerogen (vitrinite and inertinite). SEM images indicated that solid bitumen and vitrinite having same grayscale contrast. Their research indicated that solid bitumen hosted nanopores, while vitrinite stayed largely non-porous.

In the Bakken Formation, correlative microscopy revealed the challenges in distinguishing OM types using SEM only. Optical microscopy was used to identify amorphous OM, solid bitumen and inertinite. However, due to the similar appearance between those three phases, it was difficult to distinguish between the different OM phases using the SEM images. This can be a potentially fatal flaw in studies of shale reservoirs since each maceral type contributes differently in the processes of hydrocarbon generation and storage. Using both optical and SEM imaging to classify material united the two prevalent imaging techniques and kept the study from misclassifying before porosity analysis. The Eagle Ford Formation offered an almost perfect laboratory for us to investigate how the different forms of solid bitumen can occur, each with an entirely different thermal history. Optical microscopy revealed a type of high-reflectance solid bitumen post-generational that was confirmed to be rich in organic nanopores by SEM. A second type of solid bitumen, dispersed within the rock matrix and with lower reflectance, exhibited minimal porosity. These differences indicated that not all solid bitumen adds the same amount of storage space for hydrocarbons, and in the absence of employing correlative microscopy, such variations would be hard to observe.

Sethi et al. [35] in their pioneering application of OECM on Indian shales of Rajmahal and Jharia Basins observed differences in porosity between the thermally immature and mature samples. Optical microscopy was first employed for maceral classification, and later SEM imaging allowed high-resolution imaging to detail their porosity. In the thermally immature Rajmahal shale samples, liptinite was observed to have SEM visible nanopores (Fig. 3.4). In thermally mature Jharia shale samples, liptinite maceral was devoid of porosity under SEM, indicating that with advanced thermal evolution, primary pores are collapsed or occluded (Fig. 3.5). No SEM visible pores were found in the vitrinite macerals of both Rajmahal and Jharia basin shales, while only primary pores as partially preserved cell lumens were seen in semifusinite (Figs. 3.4, 3.5 and 3.6). The study highlighted that organic pore were better distinguishable at acceleration voltage of 5 kV compared to when it was set at 10 kV. This suggested that lower accelerating voltages provide better imaging

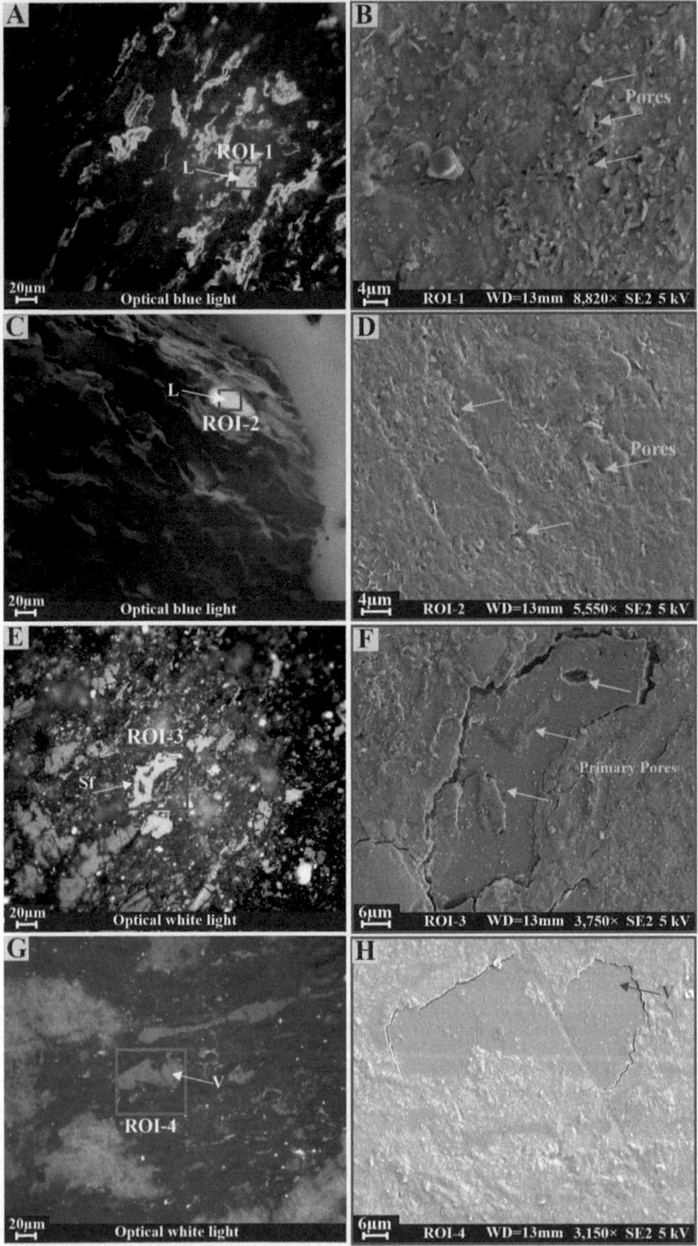

Fig. 3.4 Correlative imaging of porosity in RM shale samples: **A** Liptinite maceral identified as ROI-1; **B** SEM image showing liptinite maceral with pores **C** Liptinite maceral: ROI-2; **D** SEM image showing liptinite maceral with pores; **E** Semifusinite labeled as ROI-3; **F** SEM visualization of a mineral infillings within primary pore; **G** Vitrinite maceral designated as ROI-4; **H** SEM analysis showing no visible porosity in vitrinite Abbreviation: V—Vitrinite. (Reprinted from *International Journal of Coal Geology*, Vol. 274, Sethi, C., Mastalerz, M., Hower, J.C., Hazra, B., Singh, A.K. and Vishal, V., "Using optical-electron correlative microscopy for shales of contrasting thermal maturity," p. 104273, Copyright (2023), with permission from Elsevier. License Number: 6030310456162)

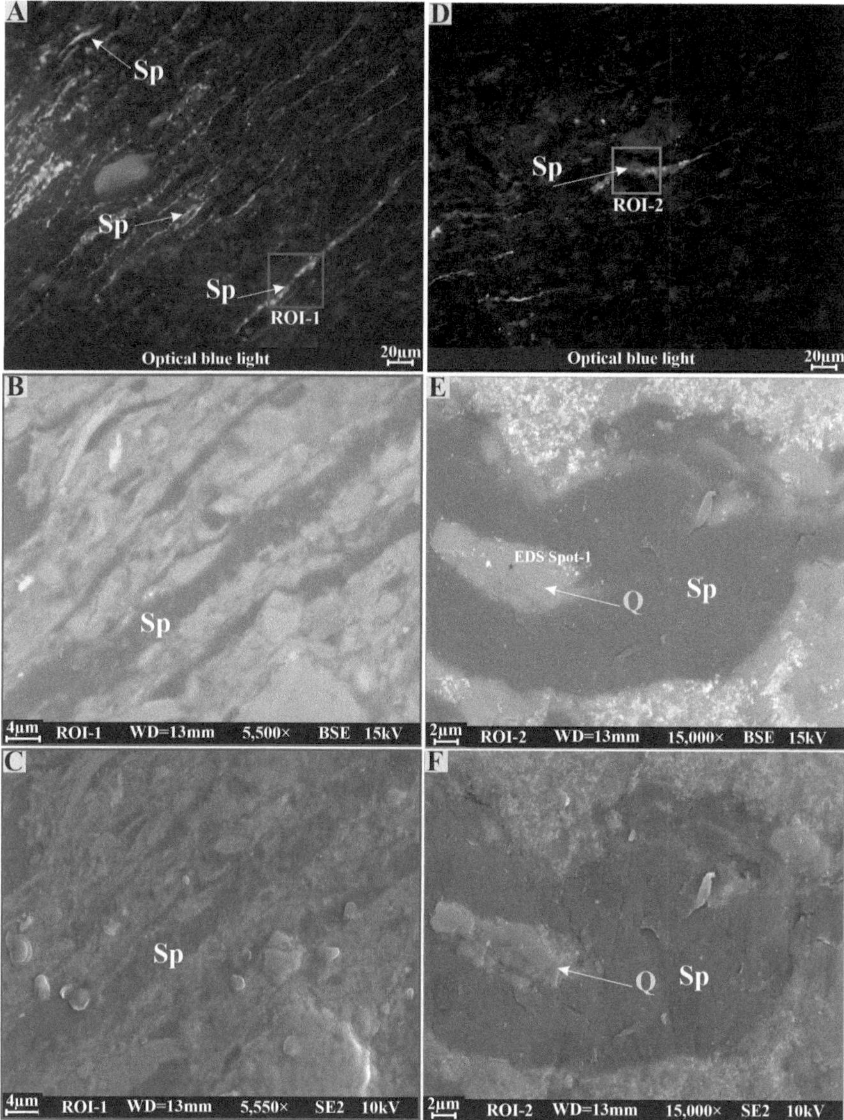

Fig. 3.5 Correlative image set of liptinite found in shale sample from Jharia Basin, India. Abbreviations: Sp- Sporinite; Q- Quartz. (Reprinted from *International Journal of Coal Geology*, Vol. 274, Sethi, C., Mastalerz, M., Hower, J.C., Hazra, B., Singh, A.K. and Vishal, V., "Using optical-electron correlative microscopy for shales of contrasting thermal maturity," p. 104273, Copyright (2023), with permission from Elsevier. License Number: 6030310456162)

conditions for detecting fine-scale porosity in OM, possibly due to reduced interaction volume effects in low-density materials like organic macerals [3].

An additional impetus of this study was the elucidation of mineral-organic associations in shales. Furthermore, the correlative microscopy approach uniquely provided the ability to accurately identify mineral matter juxtaposed within the organic matrix, which would have been nearly impossible with SEM alone. For example, energy dispersive spectroscopy (EDS) analysis showed occurrence of CMs inside the cell lumens of semifusinite macerals of Rajmahal shales as well as Jharia shales. This finding is particularly notable as it reveals novel insights into the role of mineral phases on controlling mechanical stability and thus diagenetic evolution of OM. An especially interesting part of the study was the preferred association of pyrite with vitrinite in Jharia shales. Pyrite was not easily identifiable using optical microscopy but became evident with SEM. The distribution of pyrite very near to vitrinite suggests a diagenetic relationship, perhaps even that pyrite formation may have taken place during or subsequent to maturation of OM. This close association has important consequences for hydrocarbon generation since pyrite has been shown to catalyse hydrocarbon formation, hence affecting the net petroleum potential of shales (although Ma et al. [26]). Additionally, correlative microscopy has been key in assessing the surface relief and topography of various maceral types within SEM. It was noted that liptinite macerals, especially sporinite and resinite, commonly present positive relief under SEM in thermally immature samples, but negative relief in thermally mature samples. This implies that the mechanical and chemical properties of macerals change with increasing thermal maturity, possibly due to the release of volatile compounds and corresponding maceral hardness changes. To correctly interpret these types of observations, it is critical to use complementary optical and electron microscopy to measure the physical and chemical transformations happening to OM over geological timescales.

Multi-scale imaging provides researchers better accessibility for studying complex relationships between organic and mineral phases in shales which may lead to accurate prediction of hydrocarbon generation potential of shale reservoirs. With the continued refinement of this technique its application in future research will play an important role the optimization of strategies for exploration and production of shale hydrocarbons.

3.4 Atomic Force Microscopy (AFM)-Based Characterization of Shales

AFM has recently emerged as a powerful nanoscale imaging tool for characterizing shale reservoirs, giving valuable information at the nanoscale resolution about the organic and inorganic components, pore structures and mechanical properties [19]. AFM works on the principle of scanning probe microscopy, in which a very sharp tip attached to a very flexible cantilever which scans the sample surface, measuring surface topography and material properties via atomic-scale interactions. The cantilever's deflection caused by such forces like van der Waals interaction,

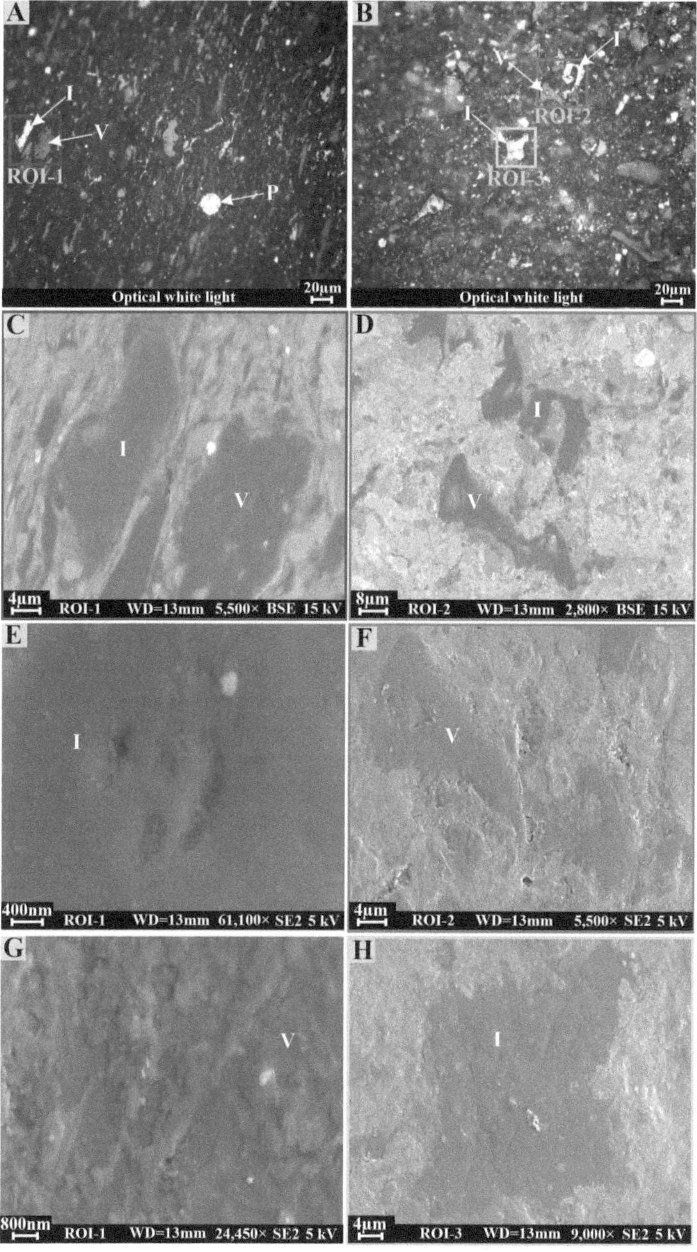

Fig. 3.6 Correlative imaging of porosity in a shale sample from Jharia Basin, India: **A** Inertinite and vitrinite macerals identified as ROI-1; **B** Additional inertinite and vitrinite macerals labeled as ROI-2 and ROI-3; **C** SEM image of inertinite and vitrinite macerals captured; **D** SEM image of inertinite and vitrinite macerals in BSE mode; **E** High-magnification SEM image of inertinite maceral; **F** SEM visualization of vitrinite maceral at 5550× magnification in SE2 mode at 5 kV; **G** Vitrinite maceral; **H** Inertinite maceral imaged. Abbreviations: I—Inertinite; WD—Working Distance; V—Vitrinite; ROI—Region of Interest; P—Pyrite. (Reprinted from *International Journal of Coal Geology*, Vol. 274, Sethi, C., Mastalerz, M., Hower, J.C., Hazra, B., Singh, A.K. and Vishal, V., "Using optical-electron correlative microscopy for shales of contrasting thermal maturity," p. 104273, Copyright (2023), with permission from Elsevier. License Number: 6030310456162)

electrostatic force, capillary force, etc. is recorded as a beam of laser reflected onto a photodetector. This feedback mechanism allows AFM to generate high-resolution 3D topographic maps of shale surfaces, revealing nanoscale heterogeneities that are otherwise undetectable by conventional imaging methods such as optical microscope or SEM [18].

3.4.1 Application of AFM for Shale Characterization

Pore structure characterization of shale using AFM

Shale formations are characterized by complex pore structure with pore sizes ranging between micro- to nano-meter scales. AFM is a useful tool for assessing the nanopore size and distribution in shales [18]. Liu et al. [24] showed that that AFM can deliver high-resolution three-dimensional topographic imaging, enabling an accurate representation of the nanopore spatial distribution. Important nanopore parameters such as pore number, areal porosity, average pore diameter, and pore size distribution was quantified using AFM. The study revealed pores with sizes ranges between 1–300 nm are suited for AFM based examination. They found that adsorption-based porosity analysis was ineffective in determining pores <4 nm due to the adsorption-induced swelling effect. Conversely, AFM was able to easily image the pore structure more reliably. Chen et al. [8] used nanometer-scale AFM with high-resolution images to characterize the surface porosity between organic-rich and organic-poor shale samples. Unexpectedly, their findings showed that organic-rich shales had much greater surface porosity, between 9.61 and 36.04%, than organic-poor shales with negligible porosity (3.18–3.43%). OM rich shales showed enhanced pore perimeter, surface area and total pore connectivity.

AFM for evaluation of micro-mechanical properties of shales

Elasticity, hardness, and adhesion study of shale through AFM can be performed by measuring the localized mechanical stiffness based on the interactions between surface of the sample and the AFM tip. AFM studies for instance have demonstrated that kerogen-rich regions are softer than surrounding mineral phases, terraces, and/ or other mechanical homogeneity implications for the mechanical behavior of shales under hydraulic fracturing. Additionally, friction force microscopy (FFM) and phase imaging techniques can be used to differentiate OM from mineral grains based on variations in surface roughness and adhesive properties [19].

Liang et al. [22] used the AFM technique and observed increase in Young's modulus of OM with increasing thermal maturity. At peak maturation, they found the modulus values showed localized decreases, due to the brittleness associated with carbonized organic material. Onumule et al. [27] found adhesive properties of OM to be affected by the composition and thermal maturity of OM. Adhesive properties of early mature OM was observed to be higher due to higher presence of C = O and hydroxyl functional groups. On the other hand, adhesive properties of OM decreased for over-mature OM due to increase in aromaticity and polar functional groups. The study highlighted progressive stiffening of kerogen with increasing thermal maturity from immature to mature stages. Lei et al. [20] observed increase in Young's modulus of OM with increasing thermal maturity, but it decreased at the overmature stage. The decrease in the overmature stage was attributed to the increase in secondary porosity. The influence of slick water on the micromechanical properties, Dai et al. [10] used AFM and found that long term interaction of shale with fracturing fluid results changes the mechanical properties due to the swelling of CM and dissolution of carbonate phases. Shale samples with abundant quartz were observed to most sound structurally, whereas the zones rich in CM were with least elasticity.

AFM-Infrared (IR) integration for shale characterization

Atomic force microscopy-based infrared spectroscopy (AFM-IR) has recently emerged as a powerful in situ technique suitable for characterizing the nanoscale chemical and mechanical properties of OM in shales. This approach enables high-resolution infrared (IR) spectra to be acquired at spatial resolutions well past the diffraction limit of conventional IR spectroscopy [1]. AFM-IR allows for chemical mapping at resolutions down to 50–100 nm. Yang et al. [44] was among the first to apply AFM-IR in Earth sciences by showing its ability to map the nano chemo-mechanical properties of solid bitumen, inertinite and Tasmanites found in artificially matured New Albany Shale samples.

Abarghani et al. [1] employed AFM-IR on thermally mature shale samples to determine chemical and mechanical heterogeneity of OM. They observed considerable reconstruction of OM due to increasing thermal maturity. Furthermore, their AFM-IR results indicated that thermally matured OM exhibited a more mechanically homogenous structure than did the immature OM which preserved a heterogeneous distribution of functional groups. It was found that, especially in the peak maturity sample, a sharp drop in fluorescence emission intensity is seen when going from the borders to the center of the region of interest (ROI). This trend suggested that the process of bituminization was further along in the central portion of the Tasmanites. Fluorescence emission indicative of the parent alginite was still present along the outer edges of the particle, as seen in Fig. 3.7B. Conceptualization of the STT framework as a mediation model over path model (Fig. 3.1 in Abarghani et al. [1]). Furthermore, the OM contained within the ROIs of both samples display considerable variation in the solid bitumen reflectance (SBRo%) values measured (Fig. 3.8C) (Fig. 3.1 in Abarghani et al. [1]). This variability presents a unique and useful opportunity to study the transitional zone between the initial alginite kerogen precursor and the secondary, petroleum-derived, solid bitumen developed in situ (Fig. 3.7D) (Fig. 3.1 in Abarghani et al. [1]).

Abarghani et al. [2] examined bacterial degradation vs thermal maturation on OM with AFM-IR. Their work investigated the specific nanoscale chemical conversions that take place within algal sourced Tasmanites and solid bitumen at early and peak maturity stages. From this, they demonstrated that OM degradation through bacterial processes created much greater chemical heterogeneity within OM, as indicated by AFM-IR spectra that exhibited substantial differences in oxygenated

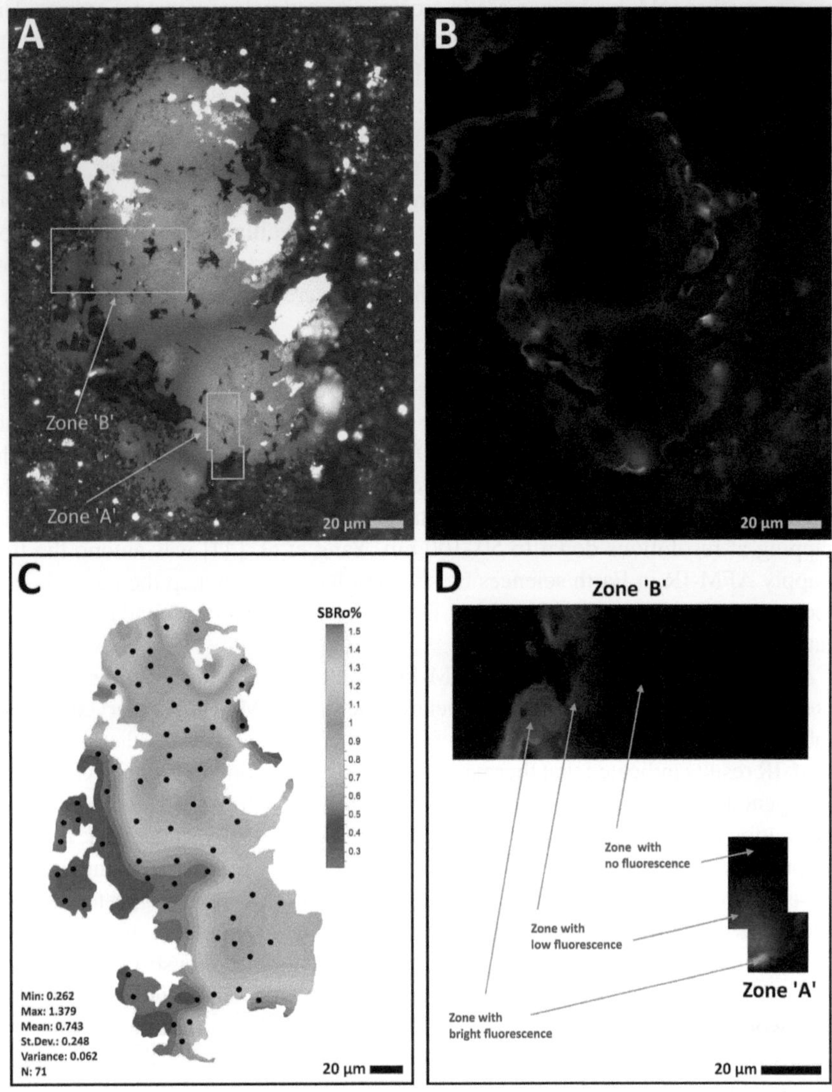

Fig. 3.7 **A** Illustration of the in-situ transformation of Tasmanites kerogen into petroleum (solid bitumen). Zones 'A' and 'B' are where the AFM-IR analysis was conducted. **B** The same field of view as in (A) but under UV light. **C** A reflectance map of SBRo% for the ROI, created using 71 measured data points represented as black dots. **D** Close-up images of the transition zones in 'A' and 'B,' showing a gradual reduction in fluorescence from the outer margin toward the center of the ROI (subzone boundaries are approximations based on fluorescence variation). **E** A panoramic topographic (height) map of zone 'B,' indicating the acquisition locations for individual spectral measurements (1–7), with red dashed lines marking approximate fluorescence subzone boundaries. **F** Zone 'B' is for Nano-IR spectra; numbers correspond to the positions in Fig. 3.8-E. **G** Zone 'A' denotes topographic map, highlighting locations of spectral data acquisition points (1–9). **H** Nano-IR spectra of zone 'A,' with numbers corresponding to the positions in Fig. 3.7-G. (Reprinted from *Geochimica et Cosmochimica Acta*, Vol. 273, Abarghani, A., Ostadhassan, M., Hackley, P.C., Pomerantz, A.E. and Nejati, S., "A chemo-mechanical snapshot of in-situ conversion of kerogen to petroleum," pp. 37–50, Copyright (2020), with permission from Elsevier. License Number: 6030320819810)

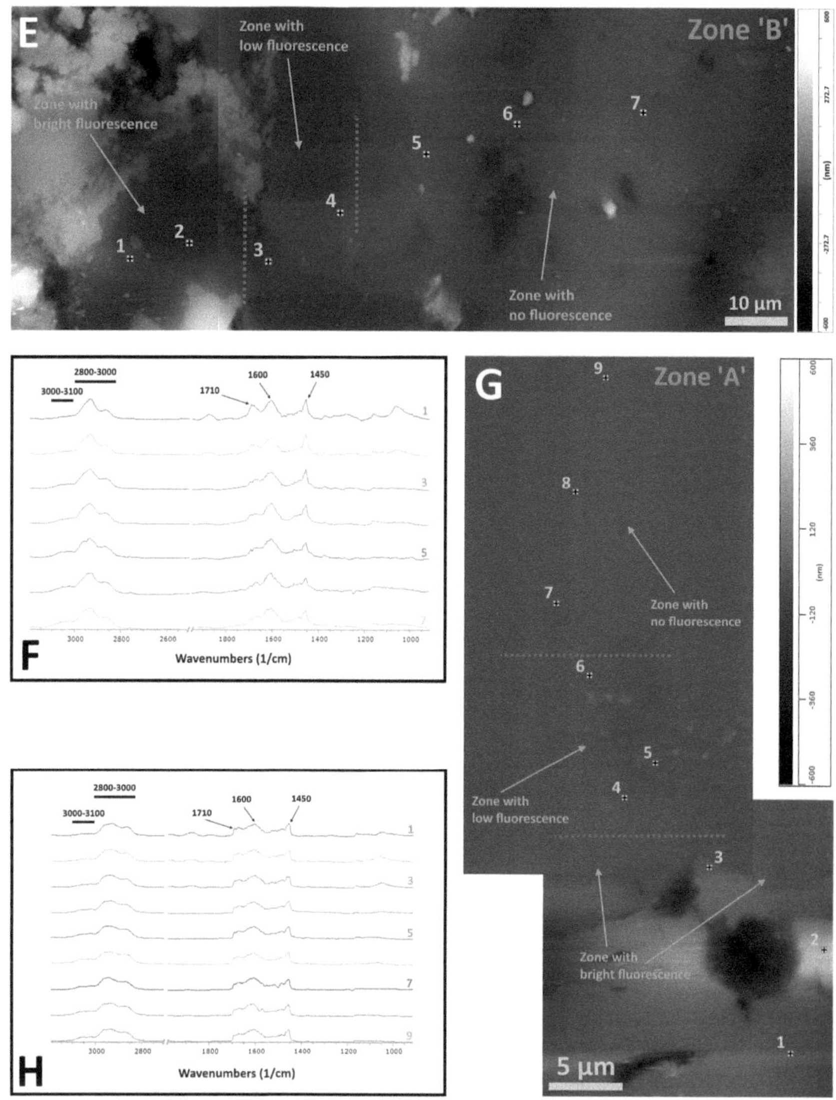

Fig. 3.7 (continued)

functional groups. Thermal maturation favoured homogenization, with Tasmanites transforming into compact solid bitumen via a progressive aliphatic compound reduction and aromaticity increase. In their study they compared unaltered telalginite with Tasmanites telalginite degraded by bacteria (kerogen Type II) side by side, separated by only a few micrometer (Fig. 3.8A and B), which provided a unique opportunity to study and compare the physicochemical impacts of both biodegradation and thermal

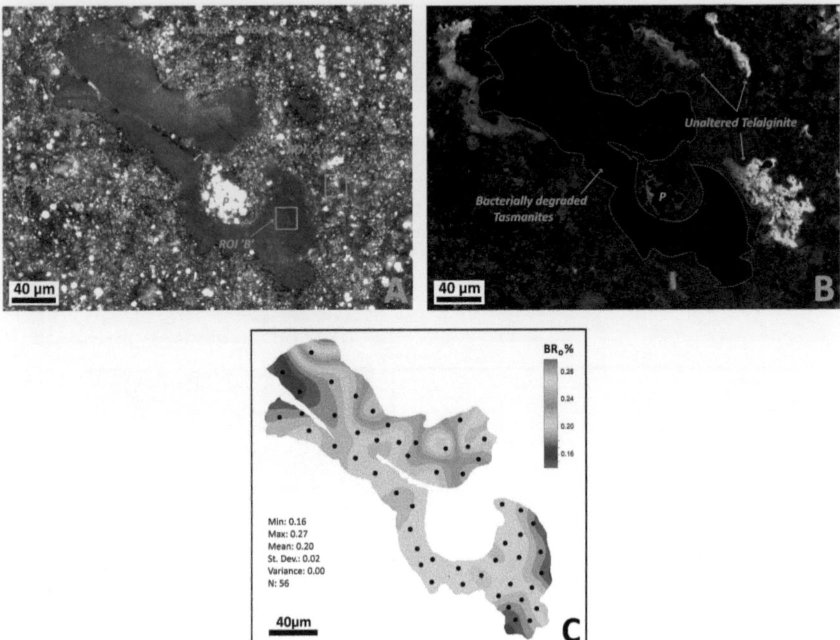

Fig. 3.8 **A** Degraded Tasmanites under white incident light identified due to its distinct morphology and texture. Framboidal pyrite (P) is concentrated at the particle's center, which is surrounded by alginite. **B** The same field of view as in (A) under UV light. The unaltered telalginite displays pale greenish-yellow to golden-yellow fluorescence, indicating an early maturity stage for this particle. (Reprinted from International Journal of Coal Geology, Vol. 223, Abarghani, A., Gentzis, T., Liu, B., Hohlbauch, S., Griffin, D., Bubach, B., Shokouhimehr, M. and Ostadhassan, M., "Bacterial vs. thermal degradation of algal matter: Analysis from a physicochemical perspective," p. 103465, Copyright (2020), with permission from Elsevier. License Number: 6030330042434)

maturation. ROI 'A' was chosen to represent the unaltered telalginite, and ROI 'B' covered part of the bacterially degraded Tasmanites. Both regions were imaged and analysed using AFM-IR to probe their nanoscale structural and chemical characteristics. Reflectance measurements made over the surface of the bacterially degraded Tasmanites (Mean BRO% = 0.20, N = 56, St. Dev. 0.05 = 0.02) were used to create an evenly spaced contour map of BRO% using a geostatistical convergent interpolation algorithm.

Zhu et al. [39] studied mirror-like surfaces of faulted shale formations and observed that slickensides on the deformed shale surfaces forced the OM to follow specific spatial patterns, with AFM-IR mapping revealing areas of graphitized carbon species concentration. These findings implied that tectonic deformation can increase the rate of organic carbon transformation, thereby influencing hydrocarbon generation and migration pathways. AFM-IR spectra demonstrated that mineral phases of quartz and clays were intimately associated with OM and played an important role in the mechanical behavior and diagenetic development of OM.

3.5   X-Ray CT for Shale Characterization

X-ray CT has emerged as key a technology for shale reservoir characterization thanks to its ability to non-destructively provide high-resolution, three-dimensional (3D) visualization of internal structures. In contrast to established, conventional, two-dimensional (2D) imaging techniques, like SEM, that solely reveal surface characteristics, X-ray CT allows researchers to investigate the three-dimensional (3D) spatial arrangement of pores, fractures and mineral phases in shale. This technique is extremely valuable and relevant for understanding shale's complex microstructural heterogeneity, which controls its porosity, permeability, and hydrocarbon storage.

X-ray CT functions from differential attenuation of X-ray beams as they pass through the sample. Depending upon the atomic composition and density of the material images of contrasting greyscale values are constructed which enables discernibility of different shale components such as OM, CM, and quartz. By adjusting the orientation of the sample while scanning, a series of angular projections can be captured that can be reconstructed with tomographic algorithms to produce high-resolution, three-dimensional models of the internal microstructure. High-density minerals (i.e., pyrite) show up very bright in X-ray CT images, while low-density components like OM and pores tend to be dark, allowing for an easy differentiation of each phase within the shale matrix [11].

One of the practical advantages of X-ray CT is its unique multi-scale characterizations of shale porosity at macro-, meso-, and micro-scales. Conventional methods of porosity measurement, like mercury intrusion capillary porosimetry (MICP), and gas adsorption, measure bulk porosity but do not have the resolution to consider spatial distribution or connectivity of pore networks [30]. X-ray CT is able to overcome these limitations by making directly visualizing the pore morphology and interconnectivity possible, an ability that is particularly important for assessing fluid transport behavior in shales. Micro-CT with a few micrometres resolution is very effective for the analysis of macropores and microfractures, whereas nano-CT with a resolution down to tens of nanometres is powerful in revealing the finer-scale porosity generally related to OM -hosted pores. Nonetheless, even with nano-CT, characterization of pores less than 50 nm is still a challenge, and integration with complementary techniques, like the focused ion beam-scanning electron microscopy (FIB-SEM), is needed for a more thorough analysis of shale pore architectures [11].

3.5.1   3D Pore System Characterization of Shales Using X-Ray CT

The primary goal that has emerged in X-ray CT-based shale characterization is the elucidation of multi-scale pore systems. Experimental studies have demonstrated that OM-hosted pores and mineral matrix pores have profoundly different geometries, connectivity and spatial distributions. High-resolution micro-CT and nano-CT imaging have revealed that organic pores tend to be highly irregular and are commonly associated with kerogen and solid bitumen, whereas mineral-associated pores often display slit-like or intergranular morphologies. Cała et al. [4] trough X-ray CT study showed that fractures in shale mainly form along the bedding planes and the mineral grain boundaries, which greatly affects the flow continuity. The visualization of fracture aperture variations and secondary mineral infillings further aids in understanding shale deformation mechanisms. Sun et al. [37] observed that X-rays are highly absorbed by high-density minerals like pyrite which makes them appear brighter, thus, are easy to identify. Mineral like quartz and carbonates are rigid which causes preservation of intergranular pores. In contrast, zones rich in clay show low porosity due to easy compaction of comparatively softer clay minerals.

Recent work using X-ray CT and focused ion beam scanning electron microscopy (FIB-SEM) by Zheng et al. [48] has shed light on fluid migration pathways in shale, showcasing multiscale characteristics of the pore system. Depending on the geometry and connectivity, pores were classified into image-processed binary images and connected pore networks yielded pore size dominated ranges of Micro-CT $= 2.759$–$5 \ \mu m$, Nano-CT $= 200$–500 nm, and FIB-SEM $= 54$–200 nm. Pore connectivity was greatest at the FIB-SEM scale (90.5%) and least at the Micro-CT scale (10.2%). Blade-shaped pores were most connected and voluminous, thus playing a major role in fluid transport. Pores showed preferential orientation parallel to the bedding planes, suggesting preferred migration pathways within shale formations at the pore scale.

The use of X-ray CT on shale studies goes far beyond static pore structure measurements. It has thus far been extensively used in dynamic experiments to study how the shale micro-structures change with different geologic and geo-engineering conditions. For example, synchrotron-based X-ray CT with high-resolution imaging capabilities has been applied to in situ shale pyrolysis observations [33], capturing formation and evolution of microfractures as kerogen thermally converted to hydrocarbons. With unique capability to conduct real-time, four-dimensional (4D, i.e., three dimensions plus time) imaging, X-ray CT is a cutting-edge technique for tracking shale reservoir evolution under complex stress, thermal, and chemical conditions [33].

3.6 Deep Learning Techniques to Segment Zones of Interest in X-Ray CT Images

Segmenting X-ray CT images of shales for detailed analysis of the inorganic and organic components of interest is desirably but challenging, and deep-learning (DL) models have been developed in recent years that are more effective for this purpose [43]. DL image segmentation techniques were pioneered for X-ray-CT-image processing for medical applications [6]. However, the use of DL techniques expanded rapidly including for use in rock-property evaluations [29]. The initial DL models, mainly based on sub-optimally configured convolutional neural networks (CNN) were computationally expensive and required substantial user-annotation (masking) of zones of interest (e.g., the Mask-R-CNN model), making them cumbersome for use with multiple-image datasets [31]. The 3D U-Net CNN model was developed to achieve semi-automated image segmentation requiring only limited user annotation for CT images of rock cores [9]. Ensemble DL models (e.g., a combination of Fast R-CNN and U-Net CNN) have also been applied to achieve automated feature detection in 2-D-CT rock-core images [28].

The relatively poor resolutions of X-ray CT images limit their applications for multi-scale analysis of pore-sized distributions in organic-rich shales. However, applying "super-resolution" techniques [45] can make it possible to create improved resolution images from low-resolution recorded CT images. These techniques have been successfully applied to distinguish pore shapes and connectivity [36]. Pixel-attention methods exploiting enhanced deep residual neural networks assists the DL methods in generating more precise representations of key input features [40]. Other CT image super-resolution techniques applied include probabilistic denoised diffusion [32] and generative-adversarial networks [23]. The diffusion models disrupt the input CT images with Gaussian noise and then use the DL models to find ways to remove that and inherent noise in the recorded images.

3.7 Convolutional DL Methods for Pore-Network Analysis of FIB-SEM Shale Images

DL models have demonstrated their effectiveness in the semantic segmentation of images for pore-network analysis [47]. Semantic segmentation divides an image into regions containing distinctive features (e.g., organic porosity, inorganic porosity, microfractures, etc.) in preference to the traditional pixel-based classification methods. The semantic segmentation approach makes it possible to establish property relationships between several distinct regions of an image. Such segmentation can be achieved using various CNN configurations.

The attention U-Net CNN model was developed to semantically segment pore-scale features of interest from rock thin section and FIB-SEM images [21]. Although the U-Net CNN model generates good results for most macropore and micropore size s it does not work well for nanopores, micro-fractures or rugged pore surfaces [46]. The limitation of the U-Net CNN is a consequence of its undercomplete network structure which steers its encoder to optimize larger more prominent features rather than low-level, smaller features such as nanopores and microfractures. To overcome this limitation, Valansarasu et al. (2021) developed a multi-branched CNN (KiU-Net CNN), which combines an overcomplete CNN component with the undercomplete

U-Net CNN. Wood [43] provides a simplified diagram of the combined model with the KiU-Net component combining 3D upsampling and MaxPooling layers in a different sequence to the U-Net component. Cross-residual fusion blocks are also incorporated to enable certain learned information to be passed in both directions between some U-Net and KiU-net layers in the combined network [41].

Yasin et al. [46] compared the ability of the KiU-Net CNN, the Attention U-Net CNN, and automated edge-threshold processing [38] to detect nanopores and microfractures in FIB-SEM images when applied to samples from five well-studied Chinese organic-rich shales (Longmaxi, Niutitang, Qianjiang Qingshankou, and Yanchang Formations) and the Bakken Shale (Canada). The image segmentation results revealed that the KiU-Net model was better able to capture nanopores in the size range 10–100 nm the other two methods in all the shales samples assessed. The KiU-Net was also able to identify smaller features, particularly those with angular shapes and irregular edges, more precisely than other DL network configurations. The KiU-Net model also converged more rapidly to optimal solutions than the Attention U-Net model, making the former a more efficient model to apply. These results suggest that the KiU-Net model has the potential to improve and automate 2-D and 3-D pore network characterization of heterogeneous, organic-rich shale formations from sizeable datasets of 2-D and 3-D FIB-SEM images.

References

1. Abarghani A, Ostadhassan M, Hackley PC, Pomerantz AE, Nejati S (2020) A chemo-mechanical snapshot of in-situ conversion of kerogen to petroleum. Geochim Cosmochim Acta 273:37–50
2. Abarghani A, Gentzis T, Liu B, Hohlbauch S, Griffin D, Bubach B, Shokouhimehr M, Ostadhassan M (2020b) Bacterial vs. thermal degradation of algal matter: analysis from a physicochemical perspective. Int J Coal Geol 223:103465
3. Belin S (1992) Application of backscattered electron imaging to the study of source rocks microtextures. Org Geochem 18(3):333–346
4. Cała M, Cyran K, Stopkowicz A, Kolano M, Szczygielski M (2016) Preliminary application of X-ray computed tomograph on characterisation of polish gas shale mechanical properties. Rock Mech Rock Eng 49:4935–4943
5. Camp WK (2016) Strategies for identifying organic matter types in SEM. In: SEPM-AAPG Hedberg research mudstone diagenesis conference, vol 16, p 19
6. Carmo D, Ribeiro J, Dertkigil S, Appenzeller S, Lotufo R, Rittner L (2022) A systematic review of automated segmentation methods and public datasets for the lung and its lobes and findings on computed tomography images. Yearb Med Inform 31(1):277–295. https://doi.org/10.1055/s-0042-1742517
7. Chandra D, Vishal V (2021) A critical review on pore to continuum scale imaging techniques for enhanced shale gas recovery. Earth Sci Rev 217:103638
8. Chen S, Li X, Chen S, Wang Y, Gong Z, Zhang Y (2021) A new application of atomic force microscopy in the characterization of pore structure and pore contribution in shale gas reservoirs. J Nat Gas Sci Eng 88:103802
9. Çiçek Ö, Abdulkadir A, Lienkamp SS, Brox T, Ronneberger O (2016) 3D U-net: learning dense volumetric segmentation from sparse annotation. In: International conference on medical image computing and computer-assisted intervention, pp 424–432. Springer, Cham

10. Dai Q, Sun W, Zuo Y, Li B, Wu Z, Tan X, Lei H, Lv W, Li Y (2025) Impact of slickwater fracturing fluid on pore structure and micromechanical properties of clay-rich shale using fluid intrusion and AFM experiments. ACS Omega

11. Gou Q, Xu S, Hao F, Yang F, Zhang B, Shu Z, Zhang A, Wang Y, Lu Y, Cheng X, Qing J (2019) Full-scale pores and micro-fractures characterization using FE-SEM, gas adsorption, nano-CT and micro-CT: a case study of the Silurian Longmaxi formation shale in the Fuling area, Sichuan Basin, China. Fuel 253:167–179

12. Hackley PC, Araujo CV, Borrego AG, Bouzinos A, Cardott BJ, Cook AC, Eble C, Flores D, Gentzis T, Gonçalves PA, Mendonça Filho JG (2015) Standardization of reflectance measurements in dispersed organic matter: results of an exercise to improve interlaboratory agreement. Mar Pet Geol 59:22–34

13. Hazra B, Singh PK, Sethi C, Pandey JK (2024) Application of optical-electron correlative microscopy for characterization of organic matter. J Geol Soc India 100(10):1385–1394

14. Hazra B, Wood DA, Mani D, Singh PK, Singh AK (2019) Evaluation of shale source rocks and reservoirs, vol 142. Springer International Publishing, Berlin, Germany

15. Hutton A (1991) Fluorescence microscopy in oil shale and coal studies

16. International Committee for Coal and Organic Petrology (ICCP) (1998) The new vitrinite classification (ICCP system 1994). Fuel 77(5):349–358

17. International Committee for Coal and Organic Petrology (ICCP) (2001) The new inertinite classification (ICCP system 1994). Fuel 80(4):459–471

18. Javadpour F (2009) Nanopores and apparent permeability of gas flow in mudrocks (shales and siltstone). J Can Pet Technol 48(08):16–21

19. Javadpour F, Moravvej Farshi M, Amrein M (2012) Atomic-force microscopy: a new tool for gas-shale characterization. J Can Pet Technol 51(04):236–243

20. Lei H, Sun W, Zuo Y, Wu Z, Dai Q, Lv W, Li Y, Xie X (2025) Impact of micromechanical properties of organic matter on the micro-mesopore structures of the over-mature shale in the Niutitang Formation. Sci Rep 15(1):4542

21. Li M, Zhang P, Hai T (2023) Pore extraction method of rock thin section based on attention U-net. J Phys Conf Ser 2467:012016. https://doi.org/10.1088/1742-6596/2467/1/012016

22. Liang Z, Jiang Z, Tang X, Chen R, Arif M (2024) Mechanical properties of shale during pyrolysis: atomic force microscopy and nano-indentation study. Int J Rock Mech Min Sci 183:105929

23. Liu B, Mastalerz M, Schieber J (2022) SEM petrography of dispersed organic matter in black shales

24. Liu X, Nie B, Wang W, Wang Z, Zhang L (2019) The use of AFM in quantitative analysis of pore characteristics in coal and coal-bearing shale. Mar Pet Geol 105:331–337

25. Liu Y, Qidi Zhang Q, Zhang N, Lv J, Gong M, Cao J (2022) Enhancement of thin-section image using super-resolution method with application to the mineral segmentation and classification in tight sandstone reservoir. J Petrol Sci Eng 216:110774. https://doi.org/10.1016/j.petrol.2022.110774

26. Ma X, Zheng J, Zheng G, Xu W, Qian Y, Xia Y, Wang Z, Wang X, Ye X (2016) Influence of pyrite on hydrocarbon generation during pyrolysis of type-III kerogen. Fuel 167:329–336

27. Onwumelu C, Kolawole O, Nordeng S, Olorode O (2024) Analyzing thermal maturity effect on shale organic matter via PeakForce quantitative nanomechanical mapping. Rock Mech Bull 3(3):100128

28. Pham C, Zhuang L, Yeom S, Shin H-S (2023) Automatic fracture characterization in CT images of rocks using an ensemble deep learning approach. Int J Rock Mech Min Sci 170:105531. https://doi.org/10.1016/j.ijrmms.2023.105531

29. Purswani P, Karpyn ZT, Enab K, Xue Y, Huang X (2020) Evaluation of image segmentation techniques for image-based rock property estimation. J Petrol Sci Eng 195:107890. https://doi.org/10.1016/j.petrol.2020.107890

30. Razavifar M, Mukhametdinova A, Nikooee E, Burukhin A, Rezaei A, Cheremisin A, Riazi M (2021) Rock porous structure characterization: a critical assessment of various state-of-the-art techniques. Transp Porous Media 136:431–456

31. Roslin A, Marsh M, Provencher B, Mitchell TR, Onederra IA, Leonardi CR (2023) Processing of micro-CT images of granodiorite rock samples using convolutional neural networks (CNN), part II: semantic segmentation using a 2.5D CNN. Miner Eng 195:108027. https://doi.org/10.1016/j.mineng.2023.108027

32. Saharia C, Ho J, Chan W, Salimans T, Fleet DJ, Norouzi M (2023) Image super-resolution via iterative refinement. IEEE Trans Pattern Anal Mach Intell 45(4):4713–4726. https://doi.org/10.1109/TPAMI.2022.3204461

33. Saif T, Lin Q, Gao Y, Al-Khulaifi Y, Marone F, Hollis D, Blunt MJ, Bijeljic B (2019) 4D In situ synchrotron X-ray tomographic microscopy and laser-based heating study of oil shale pyrolysis. Appl Energy 235:1468–1475

34. Sethi C, Hazra B, Wood DA, Singh AK (2023) Experimental protocols to determine reliable organic geochemistry and geomechanical screening criteria for shales. J Earth Syst Sci 132(2):45

35. Sethi C, Mastalerz M, Hower JC, Hazra B, Singh AK, Vishal V (2023) Using optical-electron correlative microscopy for shales of contrasting thermal maturity. Int J Coal Geol 274:104273

36. Soltanmohammadi R, Faroughi SA (2023) A comparative analysis of super-resolution techniques for enhancing micro-CT images of carbonate rocks. Appl Comput Geosci 20:100143. https://doi.org/10.1016/j.acags.2023.100143

37. Sun M, Fu J, Wang Q, Gao Z (2024) Pore network characterization and fluid occurrence of shale reservoirs: state-of-the-art and future perspectives. Adv Geo-Energy Res 12(3):161–167

38. Tian S, Bowen L, Liu B, Zeng F, Xue H, Erastova V, Chris Greenwell H, Dong Z, Zhao R, Liu J (2021) A method for automatic shale porosity quantification using an Edge-Threshold Automatic Processing (ETAP) technique. Fuel 304:121319. https://doi.org/10.1016/j.fuel.2021.121319

39. Zhu H, Lu Y, Pan Y, Qiao P, Raza A, Liu W (2024) Nanoscale mineralogy and organic structure characterization of shales: insights via AFM-IR spectroscopy. Adv Geo-Energy Res 13(3):231–236

40. Zhao H, Kong X, He J, Qiao Y, Dong C (2020) Efficient image super-resolution using pixel attention. In: Bartoli A, Fusiello A (eds) Computer vision–ECCV 2020 workshops. Lecture notes in computer science, pp 12537. Springer, Cham. https://doi.org/10.1007/978-3-030-67070-2_3

41. Valanarasu JMJ, Sindagi VA, Hacihaliloglu I, Patel VM (2021) Kiu-net: overcomplete convolutional architectures for biomedical image and volumetric segmentation. IEEE Trans Med Imaging 41(4):965–976

42. Valentine BJ, Hackley PC (2019) Applications of correlative light and electron microscopy (CLEM) to organic matter in the North American shale petroleum systems. In: Camp W, Milliken K, Taylor K, Fishman N, Hackley P, Macquaker J (eds) Mudstone diagenesis: research perspectives for shale hydrocarbon reservoirs, seals, and source rocks, vol 121. AAPG Memoir, pp 1–17

43. Wood DA (2025) Quantifying reservoir microfacies characterization using thin-section, scanned, computed tomography, and electron microscope image data. In: Implementation and interpretation of machine and deep learning to applied subsurface: problems prediction models exploiting well-log information, pp 329–360. Elsevier. https://doi.org/10.1016/B978-0-443-26510-5.00010-4. (Chapter 10)

44. Yang J, Hatcherian J, Hackley PC, Pomerantz AE (2017) Nanoscale geochemical and geomechanical characterization of organic matter in shale. Nat Commun 8(1):2179
45. Yang W, Zhang X, Tian Y, Wang W, Xue J-H, Liao Q (2019) Deep learning for single image super-resolution: a brief review. IEEE Trans Multimed 21(12):3106–3121. https://doi.org/10.1109/TMM.2019.2919431
46. Yasin Q, Liu B, Sun M, Ismail A, Wood DA, Tian X, Yan B, Fu L, Fu X (2024) An improved convolutional architecture for pore structure analysis in organic-rich shale using FIB-SEM. Int J Coal Geol 295:104625. https://doi.org/10.1016/j.coal.2024.104625
47. Yu C, Wang J, Peng C, Gao C, Yu G, Sang N (2018) Bisenet: bilateral segmentation network for real-time semantic segmentation. In: Computer vision–ECCV 2018: 15th European conference, Munich, Germany, September 8–14, 2018, proceedings, part XIII, pp 334–349. https://doi.org/10.1007/978-3-030-01261-8_20
48. Zheng H, Yang F, Guo Q, Liu K (2025) Upscaling characterizing pore connectivity, morphology and orientation of shale from nano-scale to micro-scale. Mar Pet Geol 172:107213

Chapter 4
Multi-Scale Insights into Shale Petrophysical Properties

Abstract Petrophysical data facilitates detailed multi-disciplined analysis and insights to characterize shale formations and determine their suitability for producing gas and/or oil on a commercial basis or gas storage reservoir exploitation. Mercury injection capillary pressure (MICP) and low-pressure gas adsorption (LPGA) with nitrogen (N_2) and carbon dioxide (CO_2), evaluate shale' pore-size distribution at the meso- and nano-scales. This enables the determination of pore throat and fractal dimensions, surface roughness, lacunarity, soccularity, capillary pressure and relative permeability that quantify heterogeneity and anisotropy within shale formations. Integration of MICP and nuclear magnetic resonance (NMR) pore size characteristics more comprehensively evaluates the nano-, meso- and macro-scale pore distributions. Combining MICP, NMR and LPGA analysis with field emission scanning electron microscopy (FESEM) and small-angle-neutron scattering (SANS)/ small-angle-X-ray scattering (SAXS) techniques provide more insight to nano-scale shale heterogeneities. Rock–Eval pyrolysis combined with LPGA, FESEM and optical microscopy can determine how the pore-size distribution varies with a shale's thermal maturity, type and distribution of organic matter and clay mineral content. Radar diagrams assist in the visualization of multi-dimensional characterization studies involving several shale formations. This chapter provides details of recent advancements in the execution and interpretation of the multi-scale methods described.

Keywords Mercury injection capillary pressure (MICP) · Low-pressure gas adsorption (LPGA) · Pore-size distribution · Heterogeneity/anisotropy · Multi-dimensional visualization · Field emission scanning electron microscopy (FESEM) · Small-angle-neutron scattering (SANS) · Small-angle-X-ray scattering (SAXS)

C. Sethi et al., *Recent Advances in Shale Characterization Based on Laboratory Geochemistry and Geomechanics Techniques*, SpringerBriefs in Petroleum Geoscience & Engineering, https://doi.org/10.1007/978-3-032-03961-3_4

4.1 Introduction

The characterization of petrophysical properties (porosity, permeability, and pore structure) is essential to accurately predict fluid flow, storage capacity, and mechanical behavior [14, 24]. Collectively, these properties govern the ability of the shale to both store and transport gases, affect the efficacy of hydraulic fracturing, and govern the potential success of CO_2 injection and subsequent retention [17] (Hazra et al., in press). Because of their heterogeneous and naturally multiscale pore structures, and multi-physical effects, it is still a great challenge to precisely characterize the shale petrophysics [5]. During the past decade it has become apparent [3, 29] that quantitative characterization requires multi-scale/multi-physical modelling and simulation of shale formations' key structural components. To achieve this requires multi-disciplinary analysis addressing the diverse and complex geochemical, geomechanical and petrophysical properties of organic-rich shales [20, 30, 31].

In shales, porosity can be found from the nanometer-scale micropores within organic matter (OM) and pores in mineral matrices and natural fractures [15]. Extreme permeability can be very difficult to estimate, therefore only advanced experimental or modelling techniques are able to quantify or deduce the ability of the rock to move fluids. Adsorption behavior is also controlled by pore geometry, pore connectivity, and surface characteristics [4]. The continued pursuit of energy industries in optimizing the production of hydrocarbons from unconventional resources, precise multi-scale characterization of shale has become necessary. Mercury injection capillary pressure (MICP) and low-pressure gas adsorption (LPGA) are two widely employed methods for characterization of shale petrophysical properties. Within this chapter, a comprehensive review of recent approaches and results pertaining to the petrophysical characterization of organic-rich shales is provided.

4.2 Mercury Injection Capillary Pressure (MICP)

MICP analysis provides an excellent method for determining a range of petrophysical attributes of shale, including porosity, pore throat distribution, and matrix permeability [1, 18]. These properties in tandem control the storage and movement of fluid in these low-permeability formations. The method of analysis involves injection of mercury under high pressures into the rock core [27]. As the liquid–metal can only penetrate pores when the pressure is great enough to overcome the capillary forces, the method provides an accurate profile of the size and volume distribution of pore throats. As the pressure is increased, mercury invades increasingly smaller pore, allowing for a high-resolution pore throat size distribution from macropores down to ~3 nm diameter to be obtained. This pressure–volume equilibrium relationship provides direct estimation of the free porosity available in a shale sample. The Washburn equation is frequently used to derive an inverse relationship between the pressure needed for mercury to penetrate a pore throat and the diameter of the throat

[28]:

$$r = -\frac{2\gamma \cos\theta}{P} \tag{4.1}$$

where r is the radius of the pore throat, γ is mercury's surface tension (approximately 0.48 N/m), θ is the contact angle between mercury and the rock surface (typically 140°), and P is the capillary pressure. This relationship allows a conversion of the pressure axis of MICP curve into a pore-throat-size distribution, which is important for estimating permeability and predicting capillary sealing efficiency. In the case of shale formations, notable for having a great degree of meso-pore and nano-pore complex networks, this method is extremely informative as it can penetrate a wide range of pore throat diameters (on average from 360 μm to approximately 3 nm).

Since permeability is fundamentally dictated by pore throats with the narrowest openings that create continuous, connected pathways for fluid flow, the MICP-derived pore throat size distribution is a useful surrogate for predicting matrix permeability. Methods including Swanson technique, which uses the apex of the ratio of bulk volume saturation and capillary pressure to determine the estimated permeability, have been successfully applied to MICP data [23]. Olson and Grigg [19] used the Swanson method on more than 2400 shale samples from 25 formations across 13 North American basins and found good relationships between MICP-deduced porosities and derived permeabilities. They found that MICP has the potential to be a repeatable method, not just for determining porosity, but for determining permeability when traditional testing methods are unavailable or infeasible because of lack of core material. Besides improving the confidence level of permeability predictions, MICP data would be valuable in providing more realistic reservoir quality assessment from an estimate of the gas storage potential. In shale systems, in which a large fraction of the gas is retained in nanometer-scale pores, an understanding of the pore connectivity and pore throat radius is key.

4.2.1 Recent Advances in MICP Analysis of Shales

Conventional interpretations of MICP data tend to greatly overestimate the porosity and permeability of shale samples when certain corrections are not included [6, 7]. This problem is a byproduct of the complicated behavior of shale samples when undergoing the mercury intrusion process, including mechanisms like grain compressibility, conformance effects, and compression of unreachable pores. Innovations in MICP data interpretation have improved reliability of the assessment of petrophysical properties of shales.

Recently Davudov et al. [7] proposed a more holistic correction method for porosity estimation from MICP tests on shale formations. Their method considered three main correction factors, (1) conformance, (2) compressibility of the grain, and (3) compressibility of the inaccessible pores. Conformance is the mercury needed to intrude the irregularities on the outer surface of the rock sample, prior to any true pore intrusion. This volume does not account for true porosity and needs to be subtracted from the total intrusion volume. Grain compressibility adds in the volumetric effects due to the mineral matrix changing pore volume as pressure is applied which, if uncorrected, artificially increases the calculated pore volume. Inaccessible pore compressibility is related to the deformation of very small micropores that mercury cannot access because of its relatively large molecular dimensions and high surface tension. These pores do experience a volumetric change during pressurization, albeit an unwanted source of artificially increased pore volume should be adjusted for if not corrected. Thus, a mathematical model was created to derive the compressibility of grains and pores as functions of the applied MICP test by interpreting the error-related data from the compression stage (i.e., the pressure region preceding the onset of real mercury intrusion) [7]. Their findings indicated that accessible porosity values would increase significantly (by as much as 50%) with the application of all three corrections.

Accurate petrophysical interpretation of shales also requires a thorough understanding of how intrinsic geological factors governs the evolution of a formation's pore system. Mastalerz et al. [18] used MICP to study a range of Devonian New Albany and Marcellus (U.S.A.) shale samples covering an extensive maturity range, up to advanced dry-gas thermal-maturity-window conditions (Ro ~2.5%), including some coal samples for the sake of comparison. The thermal maturity plays an important role on pore structure. Overall, the shale porosity reduced with compaction, although this was not a consistent trend. The sample IL5 (Ro 1.9%), despite being relatively mature (Ro 1.11%), showed the highest porosity and total pore volume of all the shales analysed in this study. This was mainly due to its very high quartz content, promoting the formation of interparticle pores.

The MICP injection curves produced by that study provided insight into a wide variety of capillary entry pressures, yet no clear trend between entry pressure and maturity could be found. Rather, it was the resulting pore size distribution that had a greater effect on the shape of the injection curves and the pressure needed to reach particular saturation targets. Higher volume of larger pores in samples enabled earlier mercury intrusion at lower pressures, while dominance of smaller pores needed higher pressure for intrusion. Interestingly, in their study, the authors divided the shale samples into low- and high-maturity categories and found that high-maturity shales had very few pore-throats with 10 nm or lower pore-throat diameter. This may indicate that thermal maturation lowers the abundance of the smallest pores perhaps because of the compaction or thermal transformation of OM.

A particularly novel aspect of that study was the emphasis on mercury withdrawal curves, which simulate fluid expulsion and thus serve as analogs for hydrocarbon recovery scenarios. Withdrawal efficiency, or the amount of mercury withdrawn from a sample with a decrease in pressure, was found to vary significantly from shale to shale in the study (as low as 55% and as high as 100%). This efficiency was not a function of maturity and suggested that mechanisms other than thermal transformation control retention and release of fluids. Rather, efficiency connected most directly to the distribution of certain pore size ranges, most notably, the ratio between pores sized 3–12 nm and those sized 12–50 nm. Lower ratio meant percentage of larger pores was greater than smaller pores and was related to greater withdrawal efficiency. Shale samples that possessed elevated permeability and a volumetrically more substantial fraction of pores exceeding 1000 nm in diameter were forecasted to be the most favourable precursors for early-stage gas production on account of increased interconnectivity between larger pore throats.

In the field of reservoir engineering, relative permeability curves (RP) curves play a vital role for modelling multi-phase fluid flow through a porous media. Under varying saturation conditions, the nature of flow of oil, water, or gas phases from with the rock matrix is described from RP. Historically, finding RP curves has required costly and labor-intensive lab processes including steady-state or unsteady-state core-flooding tests that are frequently complicated by factors such as core degradation, sample unavailability, and experimental error. To overcome these challenges, researchers have investigated other types of modelling approaches, such as indirect empirical models and more recently machine learning frameworks. Yet, incorporating MICP data into RP estimation has continued to be a frontier largely untapped through any deep learning approaches.

Duan et al. [8] introduced an innovative and data-efficient artificial intelligence model that connected MICP results with RP predictions. Inspired by the potential of artificial intelligence, their study developed relative permeability curve deep learning (RPCDL) model, where oil–water RP curves were directly predicted from MICP experimental data. They converted the one-dimensional MICP data (i.e., mercury injection curve, withdrawal curve, pore-throat-size distribution) into two-dimensional image matrices by using the gramian angular field (GAF) method. A GAF configuration is determined by the images generated from a time series that reflects values of pairs of variables. This transformation encoded the spatial and the temporally relative transitions of the data values, allowing for the temporal dynamics, which are often nuanced and complex, to be captured that naturally come with capillary pressure behavior. The transformed signals are expressed as RGB image representations, where each image channel corresponds to a different MICP curve. Using interpolation and normalization, the MICP signals were standardized so that all rock samples could be compared as if they were originally on the same measurement scale and pore size range. One innovative aspect of the study's methodology is the inclusion of evolutionary sequences, ordered sets of MICP image data that reproduce the gradual shift in pore structure characteristics over a set of samples. These sequences were used to train the convolutional long short-term memory (ConvLSTM) network to learn trends in the underlying rock properties that dictate relative permeability.

That study introduced self-supervised learning, using the GAF-ConvLSTM model to create tens of millions of training samples by permuting and sequencing a relatively small original dataset of only 37 core samples with paired MICP and RP data. This methodological choice proved crucial for the model's ability to learn from a sparse experimental input while retaining a strong level of generalizability. To validate the deep learning framework, a comprehensive and rigorous validation process was used. The model was trained on 37 sample size sets and tested on another 10 sample size sets. The average mean prediction error of the RP in water phase RP was 4.7% and for oil phase RP 2.5%. The strong generalization ability of the model was revealed by these results. The model was oberved to perform robustly (averaging 8.2% error on the water phase and 7.7% on the oil phase prediction) even when a tertiary set containing samples with greater variability (samples showing >20% error in porosity and permeability matching) was tested against it.

That study compared the GAF-ConvLSTM model's performance to that of standalone convolutional neural networks (CNN) and long short-term memory (LSTM) models. Their model predicted both the So and SW curves with lower median errors and fewer outliers than both of these consistently. A main advantage of the RPCDL approach is its ability to understand complex spatial–temporal relationships within MICP data that independent deep-learning models fail through static or sequential based processing. From a shale characterization standpoint, that study showed the unexplored promise of MICP data predict complex flow behavior without tedious, expensive lab testing. With the technical integration of GAF and Comvest, that type of analysis can be quantitatively accurate and physically interpretable.

4.2.2 Integration of MICP with Other Techniques for Multi-Scale Assessment

Advances in mercury intrusion interpretation methods and application of machine learning techniques have drastically increased the usefulness of MICP data for reservoir characterization. Perhaps the most telling dimension has been the use of complementary and often innovative techniques to counteract the limitations that MICP inherently carries with it. Recent advancement towards this direction involves combination of MICP and nuclear magnetic resonance (NMR) measurements.

Gu et al. [9] carried out a thorough and extensive study of eight tight sandstone and eight shale core samples from the Tuha Santanghu Basin and the Hechuan–Tongnan block (China), respectively. Their research was prompted by the often-reported phenomenon that MICP and NMR produce different pore size distributions largely because of the dynamic, pressure-driven nature of MICP compared to the static, fluid-saturated nature of NMR measurements.

In particular, MICP is biased toward overestimating mesopore volume and underestimating macropore volume as mercury must enter pores via narrow throats, a bias that is often amplified by ink-bottle shaped pore geometries. NMR, through the identification of hydrogen nuclei in brine-saturated cores, senses all connected and disconnected pore volumes and provides a more comprehensive representation of the pore body distribution. In tight sandstones, MICP underestimated the porosity contribution from large pores (>1.8 μm) and overrepresented intermediate pore sizes (0.2–1.8 μm). In shales, the limitations of MICP were even more pronounced due to the lower compressive strength of the samples, restricting the maximum injection pressures and, consequently, the detection of smaller pores (<0.0075 μm). Because NMR is sensitive to all hydrogen-bearing fluids, it showed a substantially larger percentage of micro- and nanopores, particularly in shale cores with very low permeability.

Wu et al. [33] provided an integrated multi-method study that systematically characterized petrophysical properties of source rock by employing MICP, lowfield nuclear magnetic resonance (LFNMR), and FESEM. They were aware of the fact that, contrary to the common definition of unconventional reservoirs, these materials are usually not well represented by the prevalent perception of their pore space as a single, homogeneous, porous medium, rather as a system featuring a hierarchical and heterogeneous arrangement of pore networks dictating/store flow and storage properties. To deal with this, they used a synergistic approach. MICP was used to benchmark pore throat size distributions and capillary phenomena, LFNMR produced information about the fluid-accessible pore architecture and porosity distribution as a function of hydrogen relaxation times, and FESEM image analysis along with fractal dimension analysis quantified the micrometre- and nanometre-scale heterogeneity in the morphologies of pores and fractures.

Their MICP data found distinct lithologic differences in pore throat structure. Sandstones had larger, more uniformly distributed pore throats than mudstones and limestones, which resulted in lower displacement pressures, larger mean pore throat sizes and greater overall connectivity. The mean displacement pressures for the sandstone, mudstone, and limestone during the samples tested were 7.57 MPa, 13.78 MPa, and 17.22 MPa, respectively. This was indicative of the more coarse and permeable nature of sandstone samples. The sandstone samples also demonstrated a broader range of pore throat sizes (0.01–0.16 μm), while limestone and mudstone pore systems were dominated by finer throats (mostly <0.06 μm). There was a strong negative correlation between displacement pressure and porosity for all lithologies, supporting that as porosity increases, rocks generally have a coarser pore throat and better fluid-transport characteristics.

The LFNMR characterization, complementing the MICP results, from that study further demonstrated that sandstones had higher T2-signal intensities and wider T2 distributions reflecting larger and better-connected pore networks. The T2 spectra verified that sandstone samples possessed the maximum porosity (3.02% average) next to mudstone (2.01%) and limestone (1.88%). This agreement between MICP and LFNMR datasets added to the confidence in the multi-method approach to consistently characterize the pore systems of the Taiyuan Formation samples. The study assessed pore-throat connectivity and seepage behavior using MICP derived parameters including maximum mercury saturation (Smax), residual mercury saturation (RMS), and mercury removal efficiency (MRE). Sandstone was observed to show the highest Smax and MRE, whereas limestone samples showed lowest MRE values and higher mercury saturations.

Wu et al. [33] conducted a similar study in which sandstone was again marked by bigger, more evenly spaced pores and well-defined fracture networks. Mudstones and limestones were characterized by smaller, more tortuous pore features. That study further built on this qualitative analysis by calculating fractal dimensions from the FESEM images via the box-counting method and determined fractal dimension strongly correlated with physically measurable properties including porosity and permeability. Low fractal dimension exhibiting rocks were characterized by simpler and more efficient pore systems for fluid flow. Rock that showed higher fractal dimensions were marked by pore systems which restricted fluid flow due to more complex pore geometries. This is typical of mudstones and limestones. By analysing capillary-pressure curves, that study demonstrated that although all three rock types sampled exhibited single-peaked pore-throat-distribution patterns. Sandstone samples showed consistently wider distributions to include more larger pore throats. This characteristic resulted in a comparatively greater effective permeability contribution from larger pores, while the smaller pores typical in mudstones and limestones had less of an impact on effective permeability even though they were more plentiful.

The MICP method has many advantages, it is also subject to limitations. The mercury intrusion method is inherently destructive, involves hazardous materials, and cannot characterize pores below about 3 nm. This limitation is particularly critical in organic-rich shales where a substantial volume of porosity may exist in micropores (<2 nm), which are not accessible to mercury. Therefore, for a more complete characterization of the entirety of the pore system, MICP should be used in combination with gas adsorption methods including N_2 and CO_2 low-pressure gas adsorption (LPGA) methods. N_2 adsorption is more applicable to mesopores (2–50 nm), micropores are easily characterized by CO_2 adsorption because of its smaller kinetic diameter and greater diffusivity. The combination of these techniques enables a broad multi-scale characterization of the shale pore structure.

4.3 Low Pressure Gas Adsorption Analysis (LPGA)

LPGA analysis provides a sensitive and non-destructive method for quantifying the fine-scale pore structure of shale formations [22]. This approach functions based off the idea that gas molecules, when introduced to a solid surface in a reproducible way, which is achievable under controlled pressure and temperature conditions, adsorb onto the pore walls in such a way that mimics the pore geometry. Gas adsorbed at each relative pressure is measured and valuable information regarding its petrophysical properties (volume, size and surface area) distribution can be derived. Notably, LPGA methods are especially useful for the characterization of nanopores that are below the reach of mercury intrusion, rendering them necessary for assessing pore systems under ~2 nm in width. While for characterization of mesoporous structures, the use of N_2 at cryogenic temperatures is standard, CO_2 at near-ambient temperatures has become preferred for targeting microporous domains.

Using adsorption models like the Brunauer–Emmett–Teller (BET) method along with sophisticated density functional theory (DFT) models, LPGA offers powerful, quantitative pore architecture characterizations that are vital for unraveling the physical behavior of gases within shales. By performing detailed fitting of the adsorption–desorption isotherms it is possible to deduce the degree of hysteresis, pore connectivity, and surface heterogeneity which all have a strong impact on reservoir performance. At the larger scale of multiscale shale characterization, LPGA stands as an important complement to MICP, providing critical details into pore regimes that control gas adsorption capacity and transport at the nano-scale.

4.3.1 Advances in LPGA Analysis of Shales

An important first step, prior to performing LPGA analysis, is the reduction of the sample size by crushing to increase the diffusion of gas molecules into the pore network and allow equilibrium of adsorption to reach fast. Even with broad adoption, there is still no standardized protocol with respect to the ideal particle size for these measurements. Particles less than 250 μm have been used in dozens of studies to date, the range of particle size is quite broad, ranging from coarse fragments to fine powders. These investigations bring increasing attention to the fact that such variability is not insignificant. The level of sample pulverization can have drastic effects on the results by exposing previously isolated pores or conversely by rupturing sensitive pore structures, especially within organic-rich and clay-rich matrices.

Hazra et al. [12] conducted LPGA analysis on two shale samples of contrasting thermal maturity, each crushed and sieved in to four particle size ranges (S1: 1 mm–500 μm, S2: 500–212 μm, S3: 212–75 μm and S4: 75–53 μm). Overall, their observations suggested that decreasing particle size tends to increase the accessibility of formerly inaccessible nanopores to gas molecules, especially in the smallest particle fractions. This added exposure proved to be non-linear and material-specific. In the oil-window mature shale, nitrogen adsorption indicated that even though BET surface area and pore volume both generally increased as crushing progressed to finer sizes (up to S3), crushing beyond this size (to S4) resulted in a counterintuitive reduction of BET surface area even while pore volume still increased, probably due to destruction or reconstruction of fine mesopore features. In the overmature shale, pore characteristics like surface area and pore volume showed little variation among particle sizes, suggesting a more robust pore framework that is less affected by the mechanical crushing.

Fractal dimension analysis corroborated the crush-size results of that study, demonstrating major differences as a function of particle size in the oil-mature shale but no difference in the overmature sample. Low-pressure CO_2 adsorption measurements showed that even though the finer particles (particularly S4) provided larger amounts of access to micropores less than 8 Å in diameter, they inequitably covered or destroyed a significant fraction of bigger nanopores. The study found that for very finely crushing shale samples, characterization of truly pore structure may be distorted, meaning that caution should be taken when choosing particle size for LPGA studies. The study concluded that only moderately crushed samples (between S1 and S3) gave reliable and representative porosity measurements.

He et al. [13] employed CO_2, N_2, and Ar as adsorbates to conduct adsorption analysis of Longmaxi Formation shales (China) across using five different particle size ranges. The results of the study indicated that analytical particle size has a slight impact on micropore structural parameters, including micropore volume and surface area, a major impact on mesopore and macropore structure. The micropore measurements remain relatively flat despite the strong increase in mesopore/macropore volume and surface area as the particle size decreases, as clearly seen with N_2 and Ar adsorption data (Figs. 4.1 and 4.2).

As the particle size was reduced (Figs. 4.1 and 4.2), there was a corresponding increase in average pore size (APS), mesopore and macropore volume and surface area contributions, inter-pore connectivity and aspect ratio. This trend was explained by the mechanical opening of isolated or partially closed larger pores upon crushing, providing gas molecules more access during adsorption measurements. Total pore volume (TPV) is increasingly dominated by meso- and macropores in finer particles, yet the total surface area (TSA) is still greatly contributed by micropores, which shows their continued importance even after size reduction (Fig. 4.3). Shale samples that were higher in terms of clay content showed a significant response indicating changes in mesopore and macropore parameter values. Organic-rich shales that had high quartz contents were more resistant to the mechanical alteration of their pore networks, indicating the protective role that rigid framework can play.

One of the primary recommendations from that study was the need to standardize the particle-size (crush-size) range used for LPGA analyses so that results are comparable across testing laboratories and therefore reliable. The study recommended the use of particle sizes between 250–180 μm, as this range maintains a good balance between pore accessibility and structural preservation. Specifically, they argued that sizes that are too fine, especially under 106 μm, can artificially change the pore network structure, thus creating misleading conclusions regarding shale reservoir quality.

Analysis of shale samples employing LPGA technique involves another crucial sample preparation step which included removal of adsorbed moisture in the sample by heating under vacuum. However, this process does not have a universal standard. These processing conditions are critical, as they can affect both the accessibility of the pores to large gas molecules and modify the key, natural pore architecture of thermally reactive organic matrices. The lack of methodological consistency bringing uncertainty into the interpretation of LPGA data, especially in organic-rich and immature shale systems.

Singh et al. [21] recently conducted an extensive study to evaluate the effect of degassing time and temperature on the characterization of shale pore structures obtained by LPGA. Among these, they chose two geochemically distinct shale samples, a thermally immature lignite-bearing shale (LS) originating from Palana Formation, and a thermally mature oil-window shale (BS) from the Barakar Formation (India). These samples were characterized across a matrix of degassing conditions (110, 200 and 300 °C for 3 and 12 h) with N_2 and CO_2 adsorbates to encompass mesopore and micropore features. The primary study finding was that degassing time had a lesser impact on pore characterization, while the alternative degassing temperature had a greater impact and was material dependent.

For the LS shale, which contained a larger fraction of reactive Type-II kerogen, degassing at the adopted high temperature of 110 °C maintained a characteristic

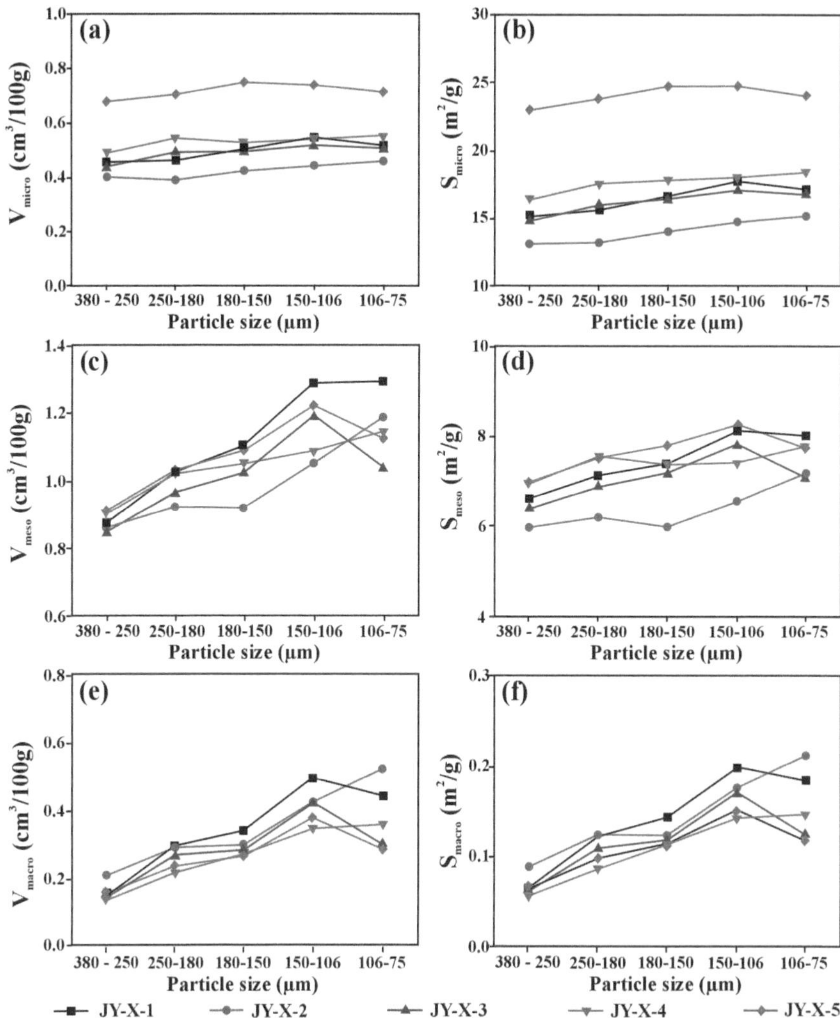

Fig. 4.1 Variation in pore volume and specific surface area of Longmaxi Formation shale samples across different particle size fractions, as determined from LPGA using CO_2 and N_2 as adsorbates. **a**, **b** Micropore volume and associated surface area; **c**, **d** mesopore volume and surface area; **e**, **f** macropore volume and corresponding surface area. (Reprinted with permission from He, Q., Dong, T., Hu, D., He, S., Yang, R. and Guo, X., "The influence of analytical particle size on the pore system measured by CO_2, N_2, and Ar adsorption experiments for shales," *Energy & Fuels*, 2021, 35(22), pp. 18637–18652. Copyright (2021) American Chemical Society)

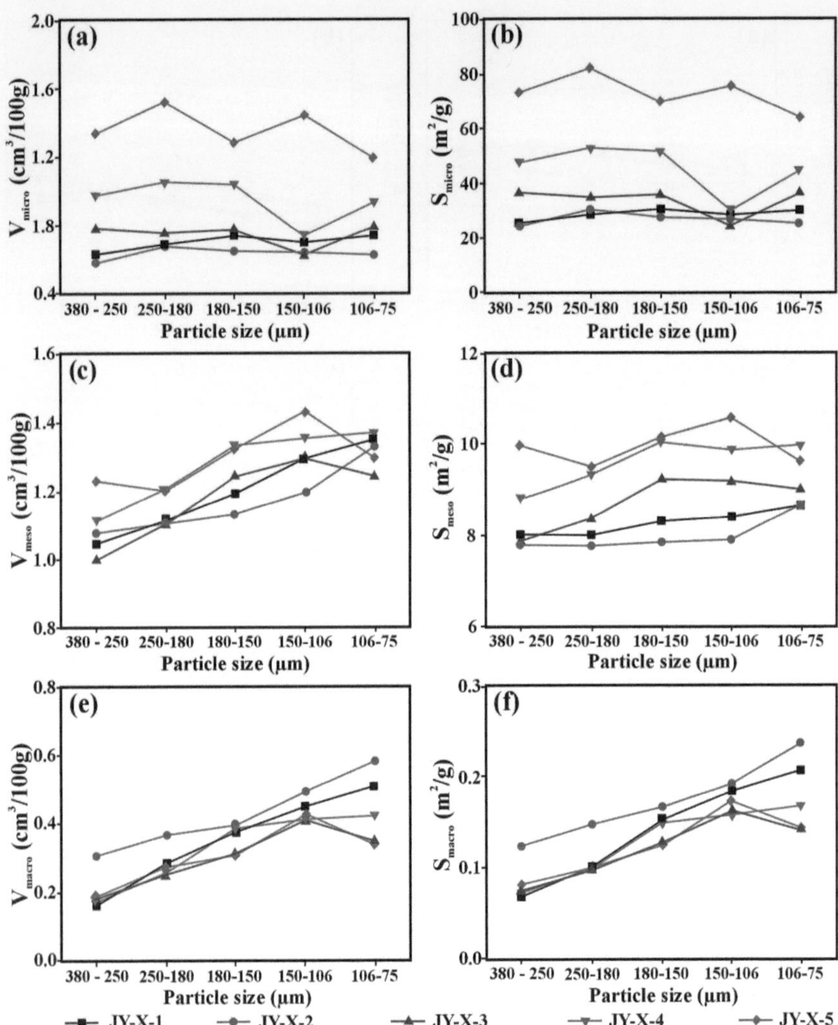

Fig. 4.2 Distribution of pore volume and surface area in Longmaxi Formation shale samples of varying particle sizes based on LPGA using argon. **a, b** Micropore volume and corresponding surface area; **c, d** mesopore volume and surface area; **e, f** macropore volume and related surface area. (Reprinted with permission from He, Q., Dong, T., Hu, D., He, S., Yang, R. and Guo, X., "The influence of analytical particle size on the pore system measured by CO_2, N_2, and Ar adsorption experiments for shales," *Energy & Fuels*, 2021, 35(22), pp. 18637–18652. Copyright (2021) American Chemical Society)

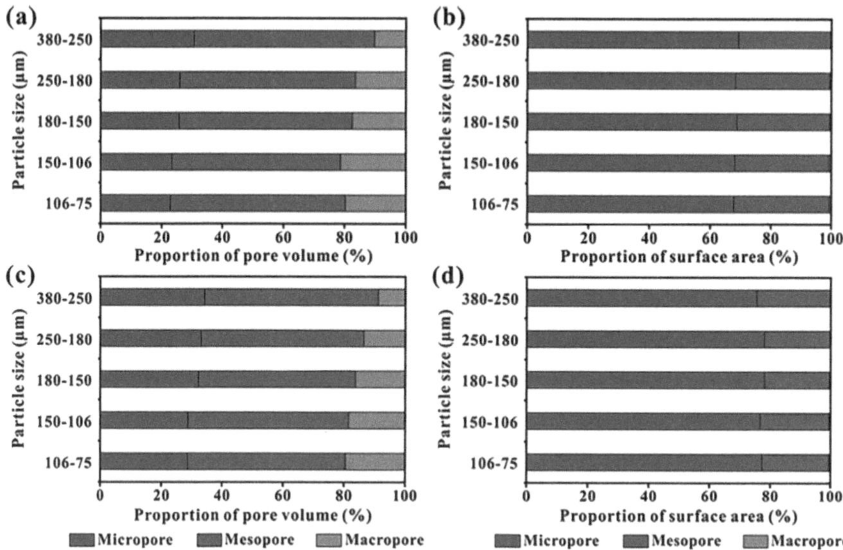

Fig. 4.3 Pore volume and surface area distribution of Longmaxi Formation shale sample obtained from combined gas adsorption measurements. **a, b** Percentage contributions of pore volume and surface area based on integrated low-pressure CO_2 and N_2 adsorption data; **c, d** percentage of pore volume and surface area derived from low-pressure argon adsorption. (Reprinted with permission from He, Q., Dong, T., Hu, D., He, S., Yang, R. and Guo, X., "The influence of analytical particle size on the pore system measured by CO_2, N_2, and Ar adsorption experiments for shales," *Energy & Fuels*, 2021, 35(22), pp. 18637–18652. Copyright (2021) American Chemical Society)

mesoporous signature, evidenced by Type IIB isotherms and narrow-mouthed ink-bottle hysteresis loops. Pore structure chemical alteration was observed at 200 °C. N_2 adsorption isotherms showed a shift toward macroporous character, with a concomitant large increase in average pore size and pore volume. This transformation was more pronounced in finer-particle fractions (<212 μm), pointing to a greater sensitivity of the pore network to thermal treatment in samples with higher surface exposure. Thermogravimetric and Rock–Eval analyses corroborated these findings by revealing the thermally reactive character of the LS shale, which experienced structural degradation and OM volatilization at high temperature. At 300 °C, even more drastic alterations were noted in the LS shale. Even though two-dimensional isotherm shapes looked similar as at 110 °C, the pore structures underneath was notably different. These results from pore-size distribution plots and fractal dimension analysis gave evidence that the pore system of LS shale, which was already partially collapsed or restructured at 200 °C, experienced additional structural rearrangement. This resulted in reduced adsorption capacities in mesopores and a proliferation of simplified pore geometries.

In contrast, a limited response to variation in degassing temperature was observed for BS shale as it was having greater thermal maturity and low abundance of reactive OM. The pore structure of BS shale was largely mesoporous at 110 °C with an intermediate surface area and volume. In contrast, a limited response to variation in degassing temperature was observed for BS shale as it was having greater thermal maturity and low abundance of reactive OM. The pore structure of BS shale was largely mesoporous at 110 °C with an intermediate surface area and volume. When degassed at 200 and 300 °C, the isotherms indicated increased macroporosity and a decline in accessible micropores (particularly for the <212 μm fraction) likely due to the formation of bitumen that blocked smaller pores. The total magnitude of change was not as extreme as the LS shale, highlighting the overshadowing role of thermal maturity and kerogen composition in dictating structural resilience.

Complementary CO_2 adsorption characterizations in that study helped to shed additional light on these phenomena within the micropore realm. Both shales showed a reduction in micropore surface area and volume after degassing at 200 °C, then some degree of recovery at 300 °C. These trends were explained by expansion of pore due to porosity generation and/or bitumen generation and volatilization which continuously modified the micropore network. DFT pore-size distribution plots showed the shifts from narrower to broader micropores, further confirming the temperature sensitivity of fine-scale porosity. Frenkel–Halsey–Hill (FHH) model-based fractal analysis further substantiated these conclusions. Among the shale samples, the LS showed the highest surface (D1) and structural complexity (D2) at 110 °C, which significantly decreased at 200 °C but revealed a partial recovery at 300 °C. The BS shale had relatively constant fractal dimensions through the temperature range, indicating that its pore geometry was not as affected by thermal simplification.

4.3.2 Integration of LPGA with Other Techniques

Due to the multiscale and heterogeneous nature of shale pore systems, no single technique can fully characterize its complex pore architecture. As an illustration, Vishal et al. [26] combined LPGA, FESEM, and SANS/SAXS in a multidisciplinary integrative approach to study the shales of the Permian Barakar Formation, belonging to the Jharia coalfield (India). The aims of that study were to establish pore size distribution (PSD), pore surface area, and to assess the ways in which these complementary methods were able to supplement one another to surpass the natural limitations of each method, thus yielding a more comprehensive characterization of the pore networks of shale formations.

Micropores and small mesopores are the major contributors to the total pore volume and surface area of N_2-LPGA. Finer pores have a higher potential for adsorption. Yet the fine resolution of the method down to sub-nanometer pores and its sensitivity to analytical models required complementary techniques. FESEM focused on direct imaging of pores and distribution within organic and mineral matrices, whereas SANS/SAXS offered views into nanoscale porosity and surface roughness at a bulk-scale level. A primary discovery from this work was that all samples showed comparable values of the fractal dimension (~2.65), meaning that pore surface textures were controlled by diagenetic processes as opposed to composition only. Another important finding of the study was that quenched surface modelling (QSDFT) produced pore-size distributions that better matched non-local DFT (NLDFT) or Barrett-Joyner-Halenda (BJH) methods.

Tian et al. [25] integrating SEM imaging, N2-LPGA, TOC and fractal analyses assessed the micro-pore PSD of shale with the inclusion of micro-fractures. They described the importance of lacunarity and succolarity for establishing a lacunarity normalizing to characterize the PSD. Lacunarity determines the space between pores in contrast to fractal dimension, which measure the irregularity of the pore surface area. Succolarity is how much of the fluid contained in the PSD can actually pass through/penetrate the PSD. As a fractal parameter, it is different than permeability, which indicates how quickly fluids can flow through the PSD, as cracking or proppant PSD permeability characterization, but rather quantifies the anisotropy of pore structures. Their findings indicated TOC content of these shale samples was positively related to the N2-adsorption capacity and nano-scale PSD. In two-dimensional space, a normalized-lacunarity metric might serve to measure PSD heterogeneity, while their computed succolarity metric might serve to identify shales prevalent with low-fluid transport capacity.

Hazra et al. [10] conducted a study to elucidate the effect of in-situ thermal alteration on the pore structure of thermally immature shale. The goal of this study was to assess the feasibility of using such treatments to improve hydrocarbon extraction and to develop CO_2-impounding microporous architectures. This work is unique in both its integrated N_2- and CO_2-LPGA with Rock–Eval pyrolysis and SEM, providing an integrated approach to characterizing pore evolution in thermally altered shales. By utilizing Langmuir porosity model, the LPGA analysis on BET data revealed intricate information about the development of mesopores and micropores with the increase in pyrolysis temperature (350, 500 and 650 °C). Figures 4.4 and 4.5 illustrate the adsorption isotherms of the raw shale sample and its thermally treated counterparts.

The N_2-LPGA derived parameters showed loss of BET surface area and increase in average pore diameter (between 350–500 °C). It was attributed to structural changes in the OM that restricted the amount of nitrogen available. In contrast, at temperatures of 500 and 600 °C, a substantial increase in CO_2-adsorption derived micropore surface area and volume was observed. A constant Type I behavior was exhibited by the CO_2 adsorption isotherms and corresponded with increased micropore capacity with 650 °C marking the ideal temperature for pore development.

Rock–Eval results found to be showing reduction in the S2 parameter and pyrolysable carbon with temperature, and a concurrent rise in the S4Tpeak (oxidation temperature of residual carbon) demonstrating advancement in the aromatisation of the OM. Complementary visual evidence from SEM imaging further bolstered this interpretation. Progressive microCT monitoring unveiled the evolution of long, organic-hosted microcracks to large, opened-up, rounded, and eventually fused pores. At 650 °C, pore wall ruptures and mergers were first observed, a sign of the formation of a more interconnected micropore network. These microstructural changes were commensurate with the used CO_2-accessible surface area and micropore volume increases.

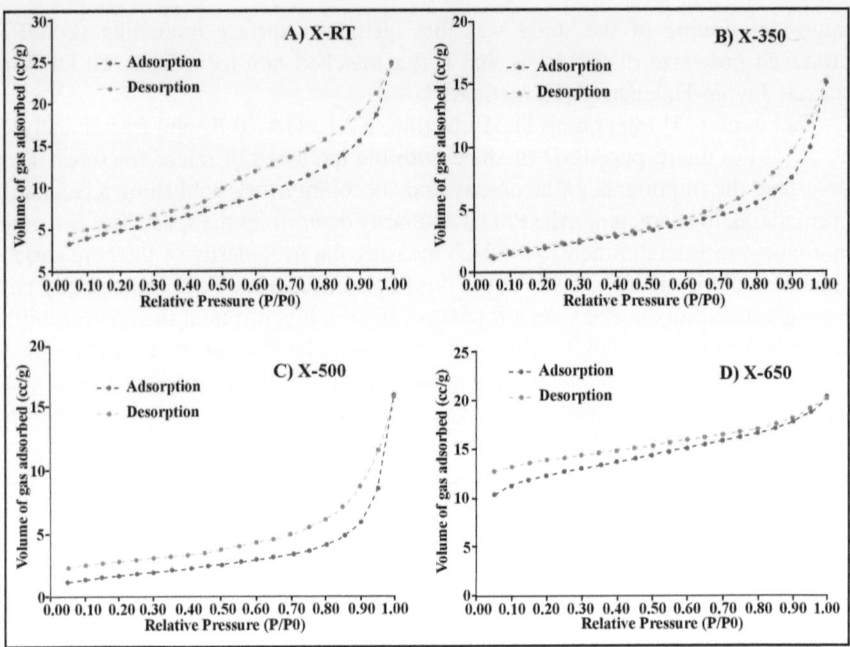

Fig. 4.4 Nitrogen gas adsorption–desorption isotherms from LPGA analysis of the Rajmahal Basin shale sample and its thermally altered variants. **A** Untreated sample (X-RT); **B** sample heated to 350 °C (X-350); **C** sample treated at 500 °C (X-500); and **D** sample exposed to 650 °C (X-650). (Reproduced from Hazra, B., Chandra, D., Vishal, V., Ostadhassan, M., Sethi, C., Saikia, B.K., Pandey, J.K., and Varma, A.K., "Experimental study on pore structure evolution of thermally treated shales: implications for CO_2 storage in underground thermally treated shale horizons," *International Journal of Coal Science & Technology*, 11(1), p. 61, 2024, Springer Nature. Published under a Creative Commons Attribution (CC BY 4.0) license http://creativecommons.org/licenses/by/4.0/)

Moisture has an important but frequently overlooked impact in the characterization of shale pore systems through gas adsorption techniques. Native water molecules that commonly exist in shale reservoirs can act as a marker that alters and strongly influences the pore accessibility measurements, as well as the inherent gas storage potential interpretation. Hydrophilic pore walls are filled by moisture (particularly those linked with clay minerals) through hydrogen bonding, that results in the creation of water films and clusters. These processes lead to blocking of pore throats and reduction of the number of available adsorption sites on internal surfaces, and changes the way adsorbate gases (e.g., CO_2) interact with the pore structure. The complexity further amplified by moisture is especially consequential under reservoir conditions, where temperature and pressure levels are above the critical point of CO_2, pushing it into a supercritical state. This key change of CO_2 creates additional complexities to the already challenging task of directly measuring adsorption capacity and pore network dynamics in shales.

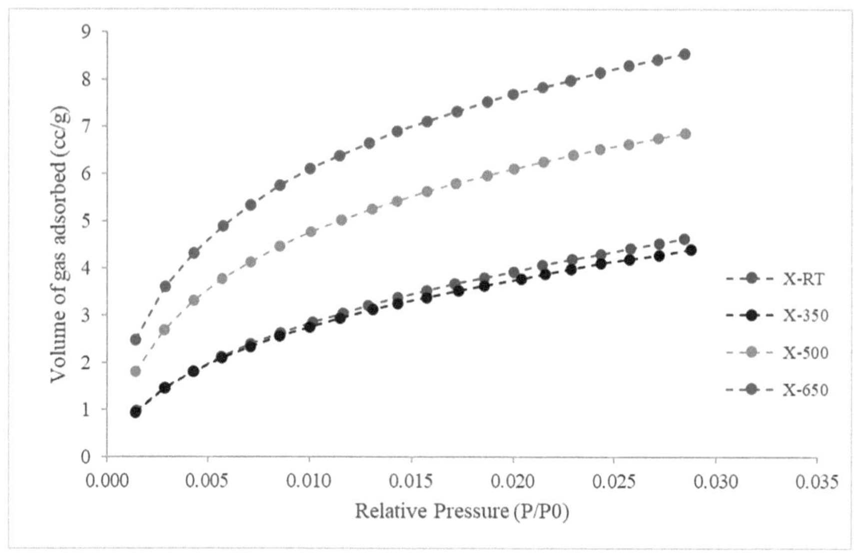

Fig. 4.5 Low-pressure CO_2 gas adsorption isotherms of the Rajmahal Basin shale sample and its thermally treated counterpart. (Reproduced from Hazra, B., Chandra, D., Vishal, V., Ostadhassan, M., Sethi, C., Saikia, B.K., Pandey, J.K., and Varma, A.K., "Experimental study on pore structure evolution of thermally treated shales: implications for CO_2 storage in underground thermally treated shale horizons," *International Journal of Coal Science & Technology*, 11(1), p. 61, 2024, Springer Nature. Published under a Creative Commons Attribution (CC BY 4.0) license http://creativecomm ons.org/licenses/by/4.0/)

Xu et al. [34] studied the impact of moisture on CO_2 adsorption and desorption behavior. Two different experimental method was used, including high-pressure CO_2 adsorption/desorption measurements up to 14 MPa at 353.15 K to simulate in situ reservoir conditions and the low-pressure CO_2 adsorption at 273.15 K to characterize micropore structures both before and after high-pressure testing. Four shale samples from the Silurian Longmaxi Formation (China), varying in mineralogical and organic composition, were studied in both dry and moisture-equilibrated conditions.

Most useful in representing change in the micropore domain were their low-pressure CO_2 adsorption isotherms at 273.15 K. These measurements showed that CO_2 adsorption/desorption can change micropore volume and surface area appreciably (more in dry samples) based on the degree of interaction between CO_2 molecules and the organic or inorganic matrix. This indicates that moisture inhibits micropore accessibility by filling fine-scale adsorption sites, decreasing the measured micropore surface area and volume. Figure 4.6 shows that these decreases in S_{micro} and V_{micro} were greatest for the dry K1 and K5 samples. Dry shale samples (K3 and K7) exhibited an increase in micropore metrics as a result of CO_2-induced structural rearrangement. Importantly, these effects differed depending on the OM and clay minerals distribution within the shale matrix, emphasizing the impacts of pore scale heterogeneity in controlling moisture influence.

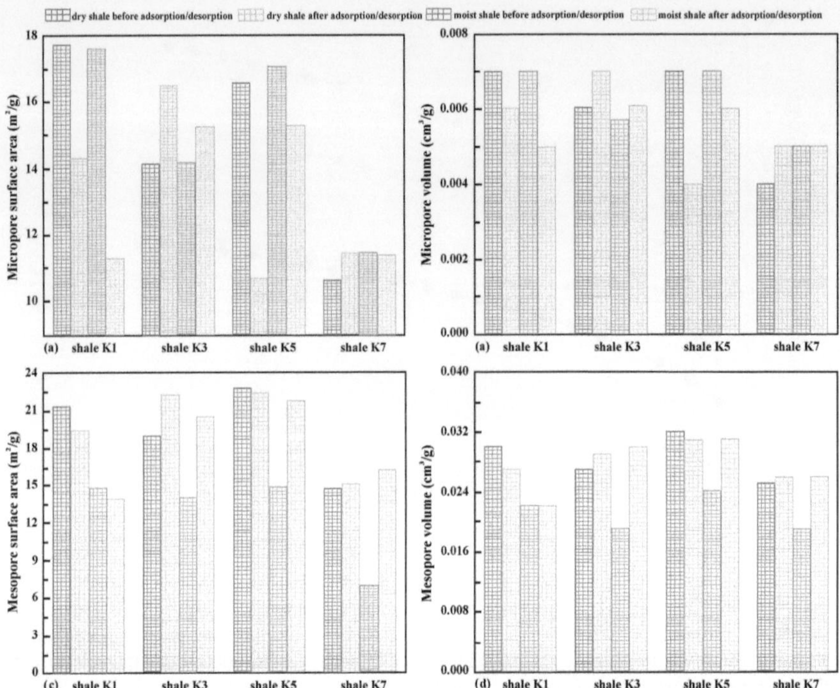

Fig. 4.6 Pore structure parameters of dry and moist Longmaxi Formation shale samples before and after adsorption/desorption. (Reprinted from *Separation and Purification Technology*, Vol. 334, Xu, Y., Lun, Z., Wang, H., Zhao, C., Zhou, X., Hu, W., Zou, J. and Zhang, D., "Understanding roles of moisture in CO_2 adsorption and desorption hysteresis on deep gas-bearing shales under high temperature and pressure," p. 125970, Copyright (2024), with permission from Elsevier. License Number: 6030290738166)

The high-pressure experiments allowed for a deeper understanding of CO_2 behavior which is more representative of realistic subsurface conditions. That study noted that dry shales showed a unique three-phase adsorption pattern with a slow increase, rapid increase and later decline in adsorption capacity with increasing pressure due to CO_2 entering its supercritical state. Moisture decreased the maximum absolute adsorption capacity by 5.78% and up to 46.39% mainly via physical blockage and competitive adsorption. Prominent hysteresis was detected in both dry and wetted shales for desorption phase showing irreversible adsorption and CO_2 trapping within the shale pore network.

Figure 4.6 also demonstrates that, in moist samples, CO_2 exposure generally results in a decrease of S_{micro} and V_{micro} values, and an associated increase in S_{meso} and V_{meso} values, except in the case of moist sample K1. This implies that moisture enhances mesopore formation through CO_2-mineral reaction while limiting micropore access.

4.4 CO$_2$-Adsorption Based Fractal Analysis of Shales

The application of fractal geometry has greatly facilitated the characterization of complex shale pore systems, especially in predicting the heterogeneity of complicated pore structures at the nano- to angstrom-scales. Among gas adsorption methods, CO$_2$-LPGA offers a distinct advantage in probing micropores below 1 nm due to the smaller kinetic diameter of CO$_2$ molecules (3.3 Å) and their strong affinity for OM (Conclusions Although Li et al. [16]). When used in conjunction with multifractal analysis, CO$_2$-LPGA allows a deeper examination of pore surface roughness, connectivity, and distribution over a wide spectrum of scales. We find this approach to be especially successful for evaluating angstrom-scale pore systems in organic rich shales, where gas storage behaviours are driven not only by pore volume but the heterogeneity and spatial complexity of the pore network.

Li et al. [16] through multifractal scaling analysis [q ~ D(q)] and singularity spectrum [α ~ f(α)] analyses revealed that all the samples showed an inverse-S shaped D(q) spectrum and parabolic convexity in f(α) plots. Such features substantiated that PSDs are extremely heterogeneous PSDs dominated by large surface asperities (unevenness), with more heterogeneity in more dense pore regions. Such features substantiated that PSDs are extremely heterogeneous PSDs dominated by large surface asperities (unevenness), with more heterogeneity in more dense pore regions.

By conducting Q-type cluster analysis on the fractal parameters, that research clustered the samples into two different shale types: Type I samples showed greater PSD heterogeneity. Type II samples had higher pore connectivity. Subsequent statistical correlation revealed that TOC contents were positively correlated with parameters representing heterogeneity (e.g., α_{-10}, D_{-10}), emphasizing the contribution of OM to the development of structurally heteromorphic micropore networks. Facies with higher dolomite content had lower heterogeneity and worse connectivity, indicating that mineralogical factors affected pore structure evolution.

Similar to organic-rich shales, coals have complex and heterogeneous PSD, especially micropores that contribute tremendously in gas adsorption (methane and carbon dioxide). Thus, the knowledge gained from using advanced pore structure technologies on coals, particularly CO$_2$-based fractal methods, can significantly aid in understanding pore networks in organic-rich shales. This is especially the case when elucidating gas storage capacity, surface heterogeneity, and pore complexity as these characteristics are just as important in SHS and for evaluating their CO$_2$ sequestration capacity.

Hazra et al. [11] focused on the CO$_2$-LPGA data on fractal dimension characterization of the microporous domain that N2-LPGA cannot sufficiently penetrate. Utilizing three independent fractal models, Dubinin–Radushkevich (DR), Frenkel–Halsey–Hill (FHH), and Mandelbrot volume-surface (V–S), the study derived pore complexity and heterogeneity metrics with precision. Overall, these fractal dimensions showed strong yet unique relationships with thermal maturity indicators and maceral compositions obtained from Rock–Eval pyrolysis and petrographic analysis. Indeed, the DR, V-S and FHH fractal dimensions obtained from CO$_2$-adsorption isotherms are in good agreement. There is still more uncertainty with the FHH-derived values since that method fits a best-fit straight line to a curved trend of data points for both coals and shales, needing more advanced curve-fitting techniques [32]. Both the DR and V-S methods are fitting straight lines to linear trends of data points.

The main conclusions of that research showed that highly mature and overmature coals exhibited much greater fractal dimensions and micropore volumes, indicative of increased surface complexity and more extensive gas storage capacity. The coal sample of medium volatile rank with rich vitrinite content, showed the maximum CO_2 adsorption capacity and micropore surface area. While the Jhama coal sample (post-mature and subjected to contact metamorphism from a neighbouring igneous intrusion) despite its extreme thermal maturity showed a much lower adsorption capacity, with this phenomenon attributable to micropore collapse under thermal stress extending to the post-mature coal, yet still retaining elevated fractal values, characteristic of residual surface irregularity.

The key methodological take-away from that study was the shown superiority of CO_2 adsorption over N_2 in determining micropore structures and their related fractal behavior. The N_2-based fractal calculations showed erratic correlations with coal maturity and maceral content. The CO_2-derived fractal dimensions reliably reflected the geochemical and petrographic evolution of the samples. This reinforces the necessity of incorporating CO_2-LPGA analysis as part of multi-scale characterization studies of organic-rich shales and coals, where microporosity and its heterogeneity fundamentally govern gas adsorption dynamics.

4.5 Comparing Multi-Scale/Multi-Physics Characteristics of Shales

One of the challenges associated with reporting quantified multi-scale/multi-physics analyses of organic-rich shales is the multi-dimensional nature of such approaches to characterize these complex formations. This is particularly the case when seeking to compare the characteristics of one shale formation with another. Graphical representations that can do this effectively need to be simple in structure, so they are relatively easy to comprehend, yet populated with quantitative data sufficient to impart the multiple numerical differences involved. Various graphical techniques could be used to achieve this.

Awan et al. [2] reported a multi-dimensional characterization study of high thermal maturity Niutitang Formation (Hunan, China) samples. That study involved Rock–Eval, bitumen reflectance, carbon-isotope, XRD, FIB-SEM, and GC-MS, analysis providing measurements related to multiple variables including TOC, thermal maturity, brittleness index, formation depth and thickness, porosity and clay content. They used a selection of six of the metrics determined to compare the characteristics of the Niutitang shale samples with published results from shale samples from ten

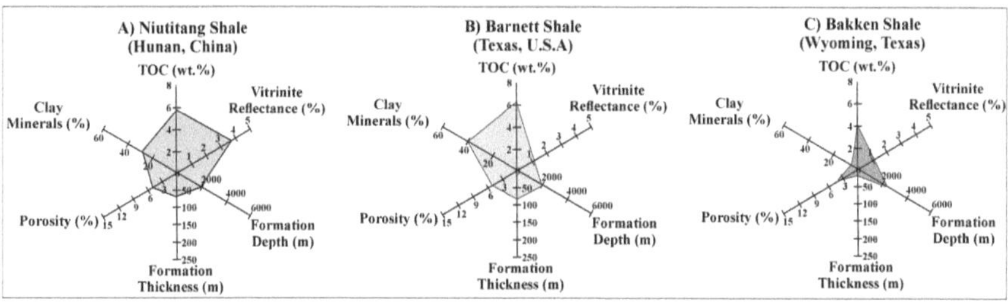

Fig. 4.7 Radar diagram displaying the analytical results of multi-physics analysis for six average metric values of three distinct shale formations. (Adapted from *Marine and Petroleum Geology*, Vol. 163, Awan, R.S., Liu, C., Yasin, Q., Liu, B., Wood, D.A., Feng, D., Wu, Y. and Iltaf, K.H., "Geologic characterization of the Lower Paleozoic black shale of the Niutitang Formation: Implications for a shale gas potential in western Hunan, China," p. 106756, Copyright (2024), with permission from Elsevier. License Number: 6030291146065)

other basins located in Europe, China, U.S.A. and Pakistan: Alum, Antrim, Bakken, Barnett, Eagle Ford, Lewis, Longmaxi, New Albany. Ohio, Talhar. That study did this effectively using a radar diagram. Figure 4.7 displays an example of that radar display including just the comparison between the average metric values for the Niutitang, Barnett and Eagle Ford shales to illustrate its conciseness and effectiveness.

Depending on the metrics analyzed in multi-scale/multi-physics characterization studies the metrics displayed on each axis of radar plots can be easily varied, making them effective graphics for comparing the results from several shale formations.

References

1. Al Hinai A, Rezaee R, Esteban L, Labani M (2014) Comparisons of pore size distribution: a case from the Western Australian gas shale formations. J Unconv Oil Gas Resour 8:1–13
2. Awan RS, Liu C, Yasin Q, Liu B, Wood DA, Dehao F, Yuping W, Iltaff KH (2024) Geologic characterization of the Lower Paleozoic black shale of the Niutitang Formation: Implications for a shale gas potential in Western Hunan, China. Mar Pet Geol 163:106756. https://doi.org/10.1016/j.marpetgeo.2024.106756
3. Aybar U, Eshkalak MO, Wood DA (2015) Advances in practical shale assessment techniques. J Nat Gas Sci Eng 27(2):399–401. https://doi.org/10.1016/j.jngse.2015.09.024
4. Bakshi T, Vishal V (2021) A review on the role of organic matter in gas adsorption in shale. Energy Fuels 35(19):15249–15264
5. Cai J, Wood DA, Hajibeygi H, Iglauer S (2022) Multiscale and multiphysics influences on fluids in unconventional reservoirs: modeling and simulation. Adv Geo-Energy Res 6(2):91–94. https://doi.org/10.46690/ager.2022.02.01
6. Comisky JT, Santiago M, McCollom B, Buddhala A, Newsham KE (2011) Sample size effects on the application of mercury injection capillary pressure for determining the storage capacity of tight gas and oil shales. In: SPE Canada unconventional resources conference?, pp SPE-149432. SPE
7. Davudov D, Moghanloo RG, Lan Y (2018) Evaluation of accessible porosity using mercury injection capillary pressure data in shale samples. Energy Fuels 32(4):4682–4694

8. Duan C, Kang B, Deng R, Zhang L, Wang L, Xu B, Zhao X, Qu J (2024) Relative permeability estimation using mercury injection capillary pressure measurements based on deep learning approaches. J Pet Explor Prod Technol 14(7):1933–1951

9. Gu Z, Wang S, Guo P, Zhao W (2024) Utilizing differences in mercury injection capillary pressure and nuclear magnetic resonance pore size distributions for enhanced rock quality evaluation: a winland-style approach with physical meaning. Appl Sci 14(5):1881

10. Hazra B, Chandra D, Vishal V, Ostadhassan M, Sethi C, Saikia BK, Pandey JK, Varma AK (2024) Experimental study on pore structure evolution of thermally treated shales: implications for CO_2 storage in underground thermally treated shale horizons. Int J Coal Sci & Technol 11(1):61

11. Hazra B, Wood DA, Panda M, Sethi C, Vishal V, Chandra D, Ostadhassan M (2025) Pore structural complexities and gas storage capacity of Indian coals with various thermal maturities. Energy & Fuels

12. Hazra B, Wood DA, Vishal V, Singh AK (2018) Pore characteristics of distinct thermally mature shales: influence of particle size on low-pressure CO_2 and N_2 adsorption. Energy Fuels 32(8):8175–8186

13. He Q, Dong T, Hu D, He S, Yang R, Guo X (2021) The influence of analytical particle size on the pore system measured by CO_2, N_2, and Ar adsorption experiments for shales. Energy Fuels 35(22):18637–18652

14. Josh M, Esteban L, Delle Piane C, Sarout J, Dewhurst DN, Clennell MB (2012) Laboratory characterisation of shale properties. J Petrol Sci Eng 88:107–124

15. Kuila U, McCarty DK, Derkowski A, Fischer TB, Topór T, Prasad M (2014) Nano-scale texture and porosity of organic matter and clay minerals in organic-rich mudrocks. Fuel 135:359–373

16. Li Z, Zhang J, Mo X, Tong Z, Wang X, Wang D, Su Z, Tang X, Gong D (2022) Characterization of an angstrom-scale pore structure in organic-rich shales by using low-pressure CO_2 adsorption and multifractal theory and its role in CH_4/CO_2 gas storage. Energy Fuels 36(19):12085–12103

17. Liang F, Zhang Q, Lu B, Chen P, Su C, Zhang Y, Liu Y (2022) Pore structure in shale tested by low pressure N_2 adsorption experiments: mechanism, geological control and application. Energies 15(13):4875

18. Mastalerz M, Drobniak A, Hower JC (2021) Controls on reservoir properties in organic-matter-rich shales: insights from MICP analysis. J Petrol Sci Eng 196:107775

19. Olson RK, Grigg MW (2008) Mercury injection capillary pressure (MICP) a useful tool for improved understanding of porosity and matrix permeability distributions in shale reservoirs. Search and Discovery Article 40322

20. Sethi C, Hazra B, Wood DA, Singh AK (2023) Experimental protocols to determine reliable organic geochemistry and geomechanical screening criteria for shales. J Earth Syst Sci 132(45):27. https://doi.org/10.1007/s12040-023-02073-6

21. Singh DP, Chandra D, Vishal V, Hazra B, Sarkar P (2021) Impact of degassing time and temperature on the estimation of pore attributes in shale. Energy Fuels 35(19):15628–15641

22. Singh DP, Hazra B, Singh V, Singh PK (2022) Review on organic porosity in shale reservoirs. In: Handbook of petroleum geoscience: exploration, characterization, and exploitation of hydrocarbon reservoirs, pp 151–171

23. Swanson BF (1981) A simple correlation between permeabilities and mercury capillary pressures. J Petrol Technol 33(12):2498–2504

24. Syed FI, AlShamsi A, Dahaghi AK, Neghabhan S (2022) Application of ML & AI to model petrophysical and geomechanical properties of shale reservoirs–a systematic literature review. Petroleum 8(2):158–166

25. Tian Z, Wei W, Zhou S, Wood DA, Cai J (2021) Experimental and fractal characterization of the microstructure of shales from Sichuan basin, China. Energy & Fuels 35(5):3899–3914. https://doi.org/10.1021/acs.energyfuels.0c04027

26. Vishal V, Chandra D, Bahadur J, Sen D, Hazra B, Mahanta B, Mani D (2019) Interpreting pore dimensions in gas shales using a combination of SEM imaging, small-angle neutron scattering, and low-pressure gas adsorption. Energy Fuels 33(6):4835–4848

27. Wardlaw NC, McKellar M (1981) Mercury porosimetry and the interpretation of pore geometry in sedimentary rocks and artificial models. Powder Technol 29(1):127–143
28. Washburn EW (1921) The dynamics of capillary flow. Phys Rev 17(3):273
29. Wood DA, Hazra B (2017) Characterization of organic-rich shales for petroleum exploration & exploitation: a review-part 1: bulk properties, multi-scale geometry and gas adsorption. J Earth Sci 28(5):739–757. https://doi.org/10.1007/s12583-017-0732-x
30. Wood DA, Hazra B (2017) Characterization of organic-rich shales for petroleum exploration & exploitation: a review-part 2: geochemistry, thermal maturity, isotopes and biomarkers. J Earth Sci 28(5):758–778. https://doi.org/10.1007/s12583-017-0733-9
31. Wood DA, Hazra B (2017) Characterization of organic-rich shales for petroleum exploration & exploitation: a review-part 3: applied geomechanics, petrophysics and reservoir modeling. J Earth Sci 28:779–803. https://doi.org/10.1007/s12583-017-0734-8
32. Wood DA (2021) Techniques used to calculate shale fractal dimensions involve uncertainties and imprecisions that require more careful consideration. Adv Geo-Energy Res 5(2):153–165. https://doi.org/10.46690/ager.2021.02.05
33. Wu D, Zhao L, Hu G, Zhang W (2025) Pore-fracture structure and fractal features of carboniferous taiyuan formation hydrocarbon source rocks as investigated using MICP, LFNMR, and FESEM. Fractal Fract 9(4):263
34. Xu Y, Lun Z, Wang H, Zhao C, Zhou X, Hu W, Zou J, Zhang D (2024) Understanding roles of moisture in CO_2 adsorption and desorption hysteresis on deep gas-bearing shales under high temperature and pressure. Sep Purif Technol 334:125970

Chapter 5
Geomechanical Characterization of Shales: Techniques and Insights

Abstract The characterization of strength and elastic properties of shales is a critical step in determining the hydrocarbon resource and extraction potential of a formation. Their inherent heterogeneity and bedding anisotropy make shales mechanically complex. Properties like mineral composition, organic content, pore structure, and thermal maturity have profound impacts on shale's mechanical behavior. Laboratory techniques like compression test, Brazilian test, and fracture toughness test are employed to measure the static mechanical properties of shales. However, these techniques require intact shale rock specimens of standard sizes which is not always available from cores recovered from wellbores due to the fissile nature of shale. Thus, micromechanical properties of shales are evaluated at finer scales using nanoindentation and atomic force microscopy techniques. These techniques reveal the mechanical heterogeneity between the organic phase and mineral phase in the shale matrix. The behavior of shales under in-situ stress and temperature conditions can be modelled using numerical simulations. Upscaling models also help in predicting the bulk-rock mechanical behavior of shales from their measured nano- and micro-scale mechanical properties. Recent advancements in machine learning techniques have enabled the use of well-log data to predict mechanical properties for large-scale reservoir assessments when core data is limited. Integration of aforementioned methods allow for efficient reservoir modelling, improved designs of wellbore fracture-stimulation programs, and safer drilling methods.

Keywords Shale geomechanics · Micromechanics · Geomechanical upscaling · Reservoir modelling · CO_2-water–rock interaction

5.1 Introduction

Geomechanical characterization of shales is an important step towards successful extraction of shale hydrocarbon resources and implementation of safe CO_2 sequestration [12]. Geomechanical properties are required for a wide range of shale exploration and exploitation activities, as summarized by Wood and Hazra [35]. Brittleness characteristics of shales have been quantified in terms of Young's Modulus (YM) and Poisson's ratio relationships for almost twenty years [23]. This has led to the recognition that the degree of anisotropy in geomechanical properties varies substantially among different shale formations, and that shale fracture permeability is substantially influenced by those properties together with effective stress, and reservoir pressure [4]. Geomechanical characteristics of shales have become essential inputs for 3-D stimulated-reservoir-volume (SRV) models used to predict the performance of multi-stage hydraulic-fracture stimulation programs, an essential tool for determining wellbore orientations, and fracture-stimulation program designs used to optimize the long-term gas or oil production from shale reservoirs [48].

At present, the most commonly used practice to increase gas and/or oil recovery from the extraction of shale resources is hydraulic fracturing, and effective fracturing operations rely heavily on the correct characterization of the geomechanical properties of shales [11, 20]. Shale formations because of their genetic complexity in terms of developed fine-scale lamination and preferred orientation of constituting minerals often lead to heterogeneous response to external loading hence affecting the failure behavior and fracability [13, 45]. Therefore, precise prediction of shale's mechanical behavior needs extensive knowledge about geomechanical parameters such as UCS, tensile strength, elastic properties (YM, Poisson's ratio, brittleness index) [24, 25].

This chapter provides a detailed summary of the geomechanical characterizations of shales, focusing on a multi-scale and multi-technique approach. It provides guidance to both established traditional static tests and new transitioning nano-mechanical procedures.

5.2 Advances in Static Mechanical Characterization of Shales

Basic understanding of the mechanical behavior of shales starts from static mechanical tests which give an insight into their strength (here defined as the envelope of failure) and deformation under various controlled stress conditions [29]. Of these, the UCS test is still the most commonly used method [15, 16]. UCS is the maximum axial stress that a rock specimen can bear under an unconstrained condition and offers essential information on wellbore stability simulation and fracturing pressure predictions [3].

Tensile strength, usually determined using Brazilian disk tests, is another important material property, particularly in terms of fracture initiation. Because of Shale's low tensile strength, a natural consequence of its easy to shear off structure with its many layers, it is very prone to crack development parallel to bedding planes. This impacts the formation of fracture through hydraulic fracturing [2]. The energy required to develop fractures within the formation is known as its fracture toughness. Its measurement is typically based on notched beam test or semi-circular bend (SCB) tests.

These mechanical parameters have been shown to be dependent on lithological composition, organic richness and pore structure. For example, a fissile shale that contains many brittle minerals such as quartz or feldspar will likely have higher UCS values as well as higher fracture toughness than a similar clay rich shale. Furthermore, thermal maturity and diagenetic history affect mechanical behavior through changes in porosity and cementation. Therefore, mechanical tests should be considered in light of the geological circumstances of the sample.

The change of rock character as fracture evolves fracturing with applied stressing can be measured in terms of stress threshold parameters [44]. The method used for measuring the strain-response of the rock under loading can directly influence the resulting stress–strain curve, thereby affecting the basis on which damage thresholds are identified. Xu et al. [43] distinguished two stress-threshold measurement methods: (i) displacement sensors known as extensometers, and (ii) linear variable differential transformer (LVDT). That study identified the impact of those two axial-strain measurement methods on the damage threshold evolution by conducting compression tests under confining conditions using Longmaxi shale (China) samples with seven distinct bedding plane orientation. They observed that the crack-initiation stress measured for the samples yielded similar results using LVDT and extensometer. However, the crack-damage stress and its corresponding strain threshold were lower when measured with the extensometer. Those differences were attributed to the limited range of the extensometer that only focuses on the central region of the specimen leading to an underrepresentation of global deformation.

Xu et al. [43] also conducted a numerical simulation using the discrete element method (DEM). The more localized strain measurements of the extensometer resulted in higher elastic modulus and Poisson's ratio values compared to those derived using a LVDT. That study also found that there was a difference between the stress–strain curves derived from the two mentioned measurement methods. The post-peak stage of axial strain measured using the extensometer showed a decreasing trend which is the characteristics of a Class II curve (Fig. 5.1a). On the other hand, LVDT-based axial strain continued to increase after the post-peak stage showing the characteristics of a Class I curve (Fig. 5.1a). A similar phenomenon was also observed during circumferential deformation load control mode (Fig. 5.1b).

Temperature's impact on the mechanical properties of rocks has been extensively studied in the recent decade. In general, the temperature-induced impact on the rock properties is studied after the thermal treatment of the rock samples. However, the impact of heat due to real-time heating of the rocks has been less explored. Wang et al. [32] conducted UCS tests of shale samples that were temperature treated (25–500 °C) under real-time condition i.e., the samples were heated during the compression tests. The compressive strength and elastic modulus of the rock was lowest at 400 °C

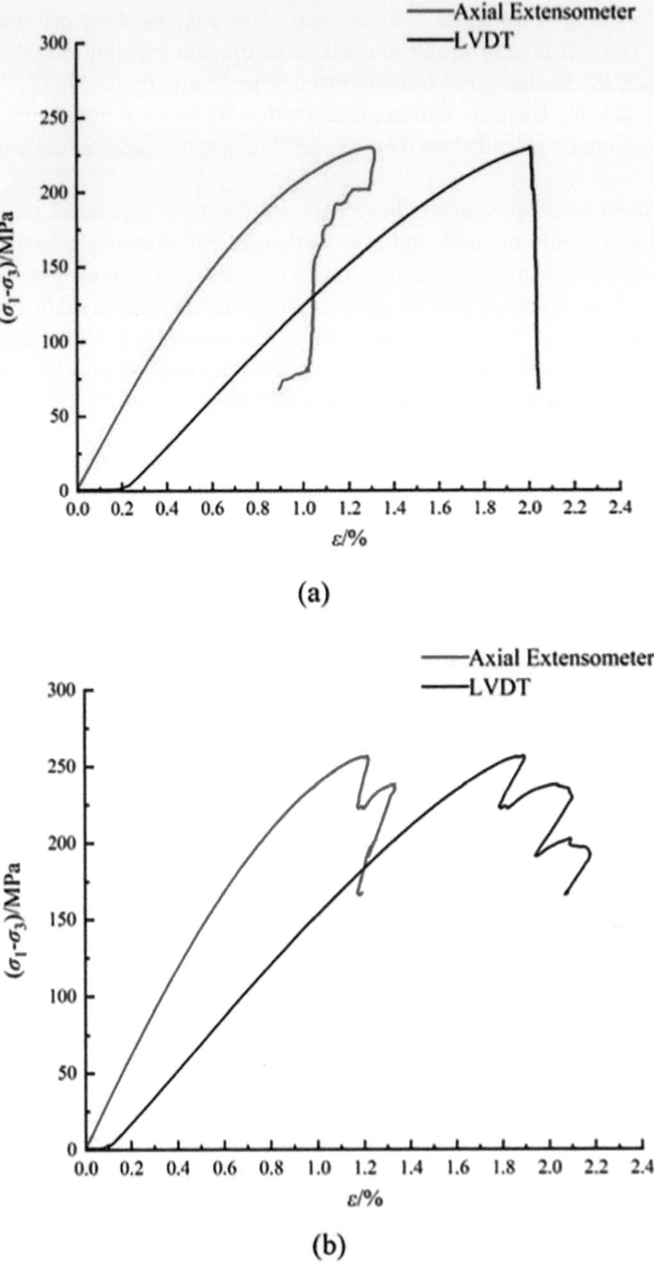

Fig. 5.1 Stress–strain curves as observed under different load control modes: load control mode involving **a** axial deformation; **b** circumferential deformation. Axes: $(\sigma_1-\sigma_3)$: deviatoric stress; ε: strain. (Reprinted from Xu, R., Gao, L., Jin, Y., Wang, Y., Li, Z. and Hu, S., "Influence of strain measurement methods on crack initiation and crack damage thresholds of shale," *Geomechanics and Geophysics for Geo-Energy and Geo-Resources*, Vol. 11(1), p. 34, Copyright (2025), with permission from Springer Nature. License Number: 6030270234361)

(Fig. 5.2). However, at temperatures above 400 °C the pyrolysis induced mineral transformations reversed the downward trends as the conversion of kaolinite to meta-kaolinite began to reinforce the shale matrix. Moreover, when loaded perpendicular to the bedding plane, the shale samples showed brittle failure followed by semi-brittle and plastic stages but then returned to brittle behavior at higher temperatures. In contrast, for samples loaded parallel to the bedding plane, plastic deformation ensued progressively due to fracture development along bedding planes. That study attributed this behavior to significant changes in mineralogy at temperatures above 400 °C.

Feng et al. [9] carried out high-rate dynamic split Hopkinson pressure bar (SPHB) tests with high-speed digital image correlation (DIC) technique to analyze tensile behavior of shale samples. High-strain rates caused instant failure and increased tensile strength of the shale samples with development of cracks that were more pronounced. Furthermore, the cracks were observed to form symmetrically along the loading direction at 0° and 90° bedding orientation, whereas crack development became increasingly complex at loading angles between 15° and 75°. The study

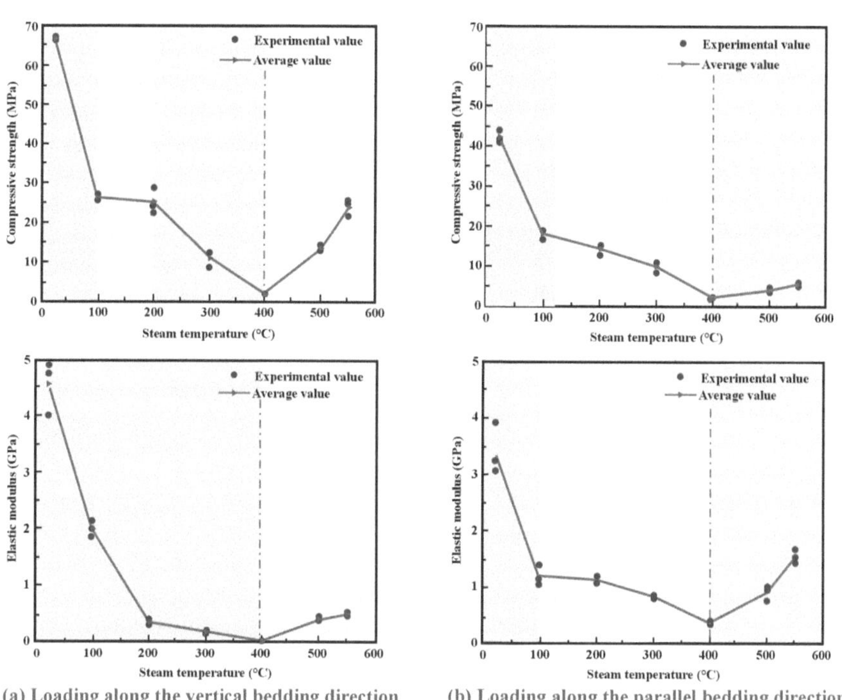

Fig. 5.2 Evolution of UCS and elastic modulus of oil shale under real-time heating. (Reprinted from Wang, L., Wang, G., Yang, D., Zhao, J., Kang, Z., Zeng, Q. and Zhao, Y., "Research on the mechanical properties and anisotropy evolution of uniaxial compression of oil shale under real-time high-temperature steam," Rock Mechanics and Rock Engineering, Vol. 58(2), pp. 1911–1932, Copyright (2025), with permission from Springer Nature. License Number: 6030260501725)

included Nova-Zaninetti criterion modified for incorporating the impact of strain rate and anisotropy into a singular strength prediction model. The modified model was observed to provide significantly improved prediction over existing models such as Hobbs-Barron and the single plane of weakness (SPW) models.

Understanding shale's elastic anisotropy under in-situ subsurface stress conditions is important for improving reservoir modelling and for developing optimization of strategies for resource extraction from these formations. However, most studies use cylindrical core samples for uniaxial or triaxial testing which do not replicate the true in-situ states of stress. Sethi et al. [28] compared elastic anisotropy of two shale lithotypes (Grey Shale and Silty Shale) under isostatic pressure and rising temperature conditions using a true-triaxial apparatus. Increase in pressure was observe to cause rise in both P-wave and S-wave velocities, whereas the velocities decreased with increasing temperature (Figs. 5.3 and 5.5). Grey shale samples showed higher velocities compared to silty shale samples due to quartz content being more in grey shale samples, whereas clay minerals was predominant in silty shale samples.

The Grey shale samples exhibited less pronounced shear-wave splitting, whereas the Silty shale displayed more pronounced shear-wave splitting under increasing pressure and temperature conditions (Fig. 5.4 and 5.6). That study also quantified static and dynamic elastic moduli and found that both shale lithotypes showed an increase in static and dynamic elastic moduli as pressure increased and decreased with increasing temperature (Fig. 5.7). The elastic anisotropy of shale, calculated from Thomsen parameters (ε and γ), showed that ε and γ parameters both decreased with increasing pressure, where Silty shale samples had a larger reduction in anisotropy than the Grey shale samples.

5.3 Numerical Modelling for Shale Mechanical Characterization

Accurate modelling of the response of rock formations to pressure is important to determine the stability of underground geological structure for reservoir engineering purposes. Models are required as direct in-situ samples of formations (e.g., by recovering cores) to determine static mechanical properties may not be available. Additionally, laboratory-based measurement of the mechanical properties of rocks never completely captures the anisotropic complexity presented by shale. Wang et al. [34] used an advanced simulation method that included a dual-model calibration framework integrating the Parallel Bond Model (PBM) and the Smooth Joint Model (SJM). They calibrated the mesoscopic parameters to reproduce the macroscopic behavior of shale specimen. Their research needed extensive, purposeful and systematic parametric studies, as well as the fitting of laboratory data on occasion. This made possible to estimate the PBM main parameters as particle elastic modulus

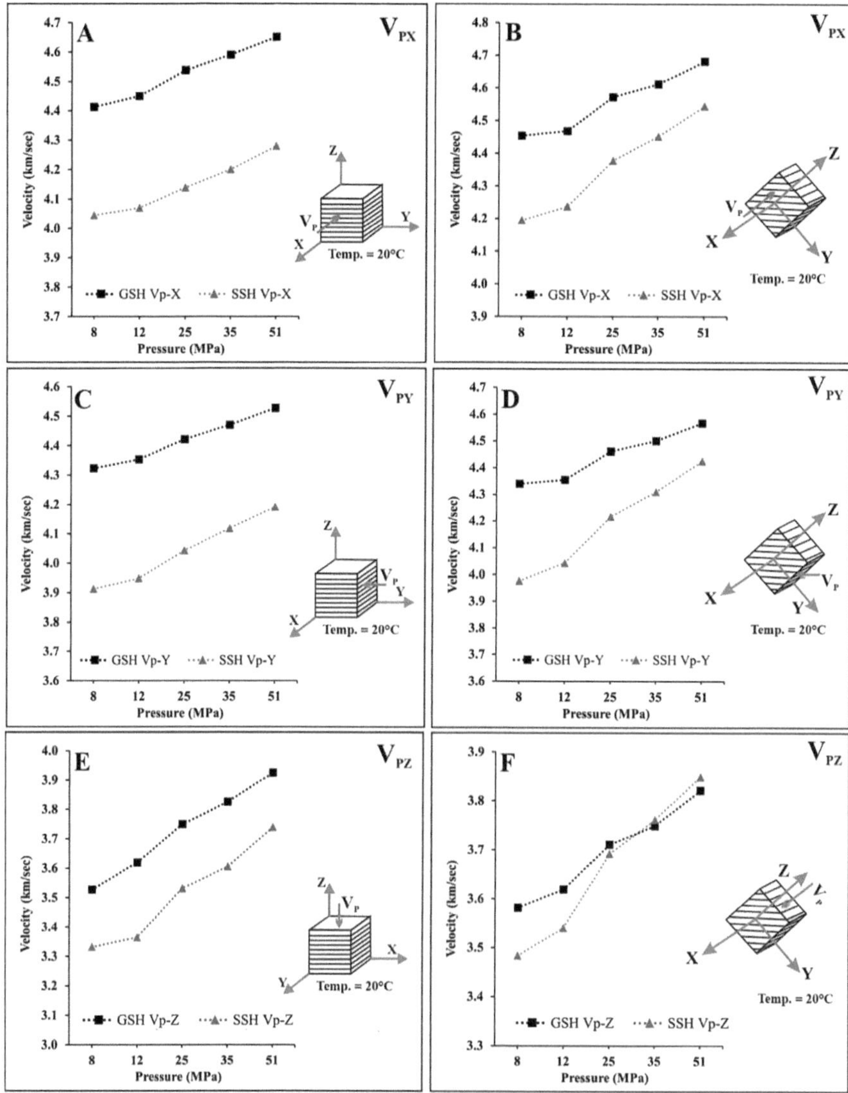

Fig. 5.3 Variation in P-wave velocity (V_P) in all three directions (parallel, perpendicular, and 45°) with respect to the bedding plane with increasing isostatic pressure. Abbreviations: GSH- Grey Shale; SSH- Silty Shale. (Reproduced from Sethi, C., Motra, H.B., Hazra, B. and Ostadhassan, M., "Influence of lithological contrast on elastic anisotropy of shales under true-triaxial stress and thermal conditions," *International Journal of Rock Mechanics and Mining Sciences*, Vol. 190, p. 106100, Copyright (2025), published under a Creative Commons Attribution (CC BY 4.0) license https://creativecommons.org/licenses/by/4.0/)

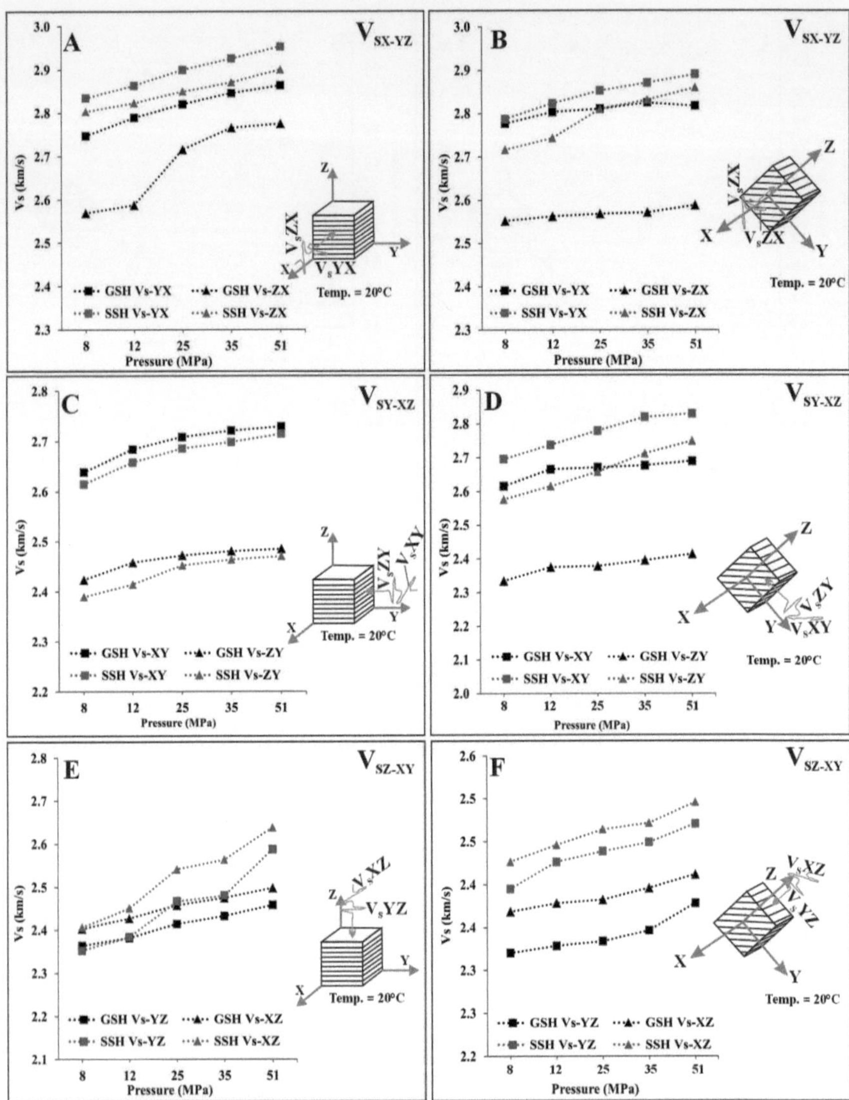

Fig. 5.4 Variation in S-wave velocity (V_S) in in all three directions (parallel, perpendicular, and 45°) with respect to the bedding plane with increasing isostatic pressure. Abbreviations: GSH- Grey Shale; SSH- Silty Shale. (Reproduced from Sethi, C., Motra, H.B., Hazra, B. and Ostadhassan, M., "Influence of lithological contrast on elastic anisotropy of shales under true-triaxial stress and thermal conditions," *International Journal of Rock Mechanics and Mining Sciences*, Vol. 190, p. 106100, Copyright (2025), published under a Creative Commons Attribution (CC BY 4.0) license https://creativecommons.org/licenses/by/4.0/)

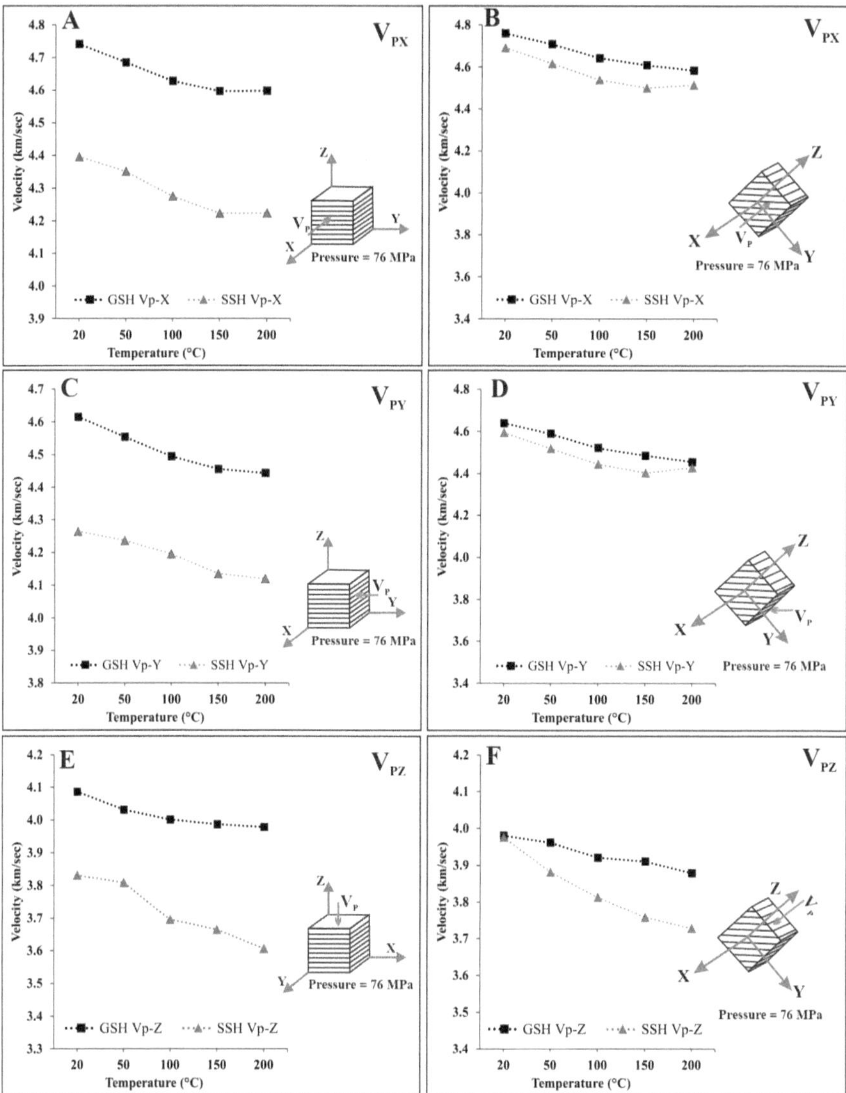

Fig. 5.5 Variation in P-wave velocity (V_P) in in all three directions (parallel, perpendicular, and 45°) with respect to the bedding plane with increasing temperature. Abbreviations: GSH- Grey Shale; SSH- Silty Shale. (Reproduced from Sethi, C., Motra, H.B., Hazra, B. and Ostadhassan, M., "Influence of lithological contrast on elastic anisotropy of shales under true-triaxial stress and thermal conditions," *International Journal of Rock Mechanics and Mining Sciences*, Vol. 190, p. 106100, Copyright (2025), published under a Creative Commons Attribution (CC BY 4.0) license https://creativecommons.org/licenses/by/4.0/)

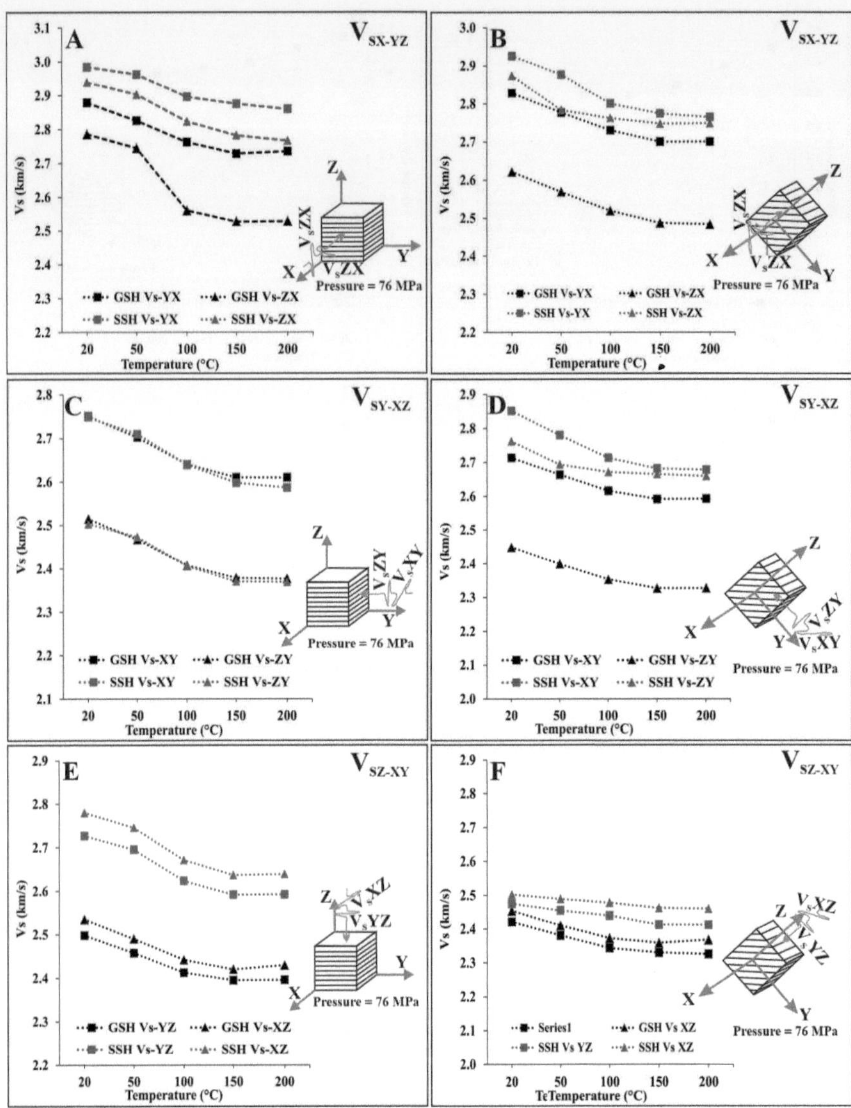

Fig. 5.6 Variation in S-wave velocity (V_S) in in all three directions (parallel, perpendicular, and 45°) with respect to the bedding plane with increasing temperature. Abbreviations: GSH- Grey Shale; SSH- Silty Shale. (Reproduced from Sethi, C., Motra, H.B., Hazra, B. and Ostadhassan, M., "Influence of lithological contrast on elastic anisotropy of shales under true-triaxial stress and thermal conditions," *International Journal of Rock Mechanics and Mining Sciences*, Vol. 190, p. 106100, Copyright (2025), published under a Creative Commons Attribution (CC BY 4.0) license https://creativecommons.org/licenses/by/4.0/)

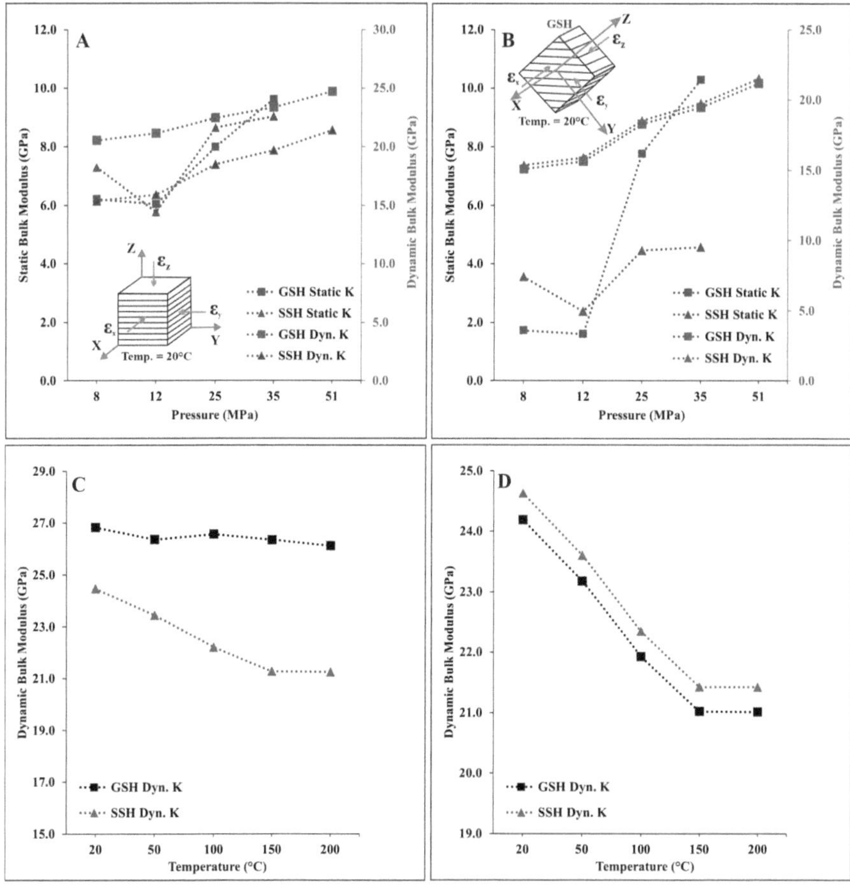

Fig. 5.7 Variations in static and dynamic modulus of distinct shale lithotypes. Abbreviations: GSH-Grey Shale; SSH- Silty Shale. (Reproduced from Sethi, C., Motra, H.B., Hazra, B. and Ostadhassan, M., "Influence of lithological contrast on elastic anisotropy of shales under true-triaxial stress and thermal conditions," *International Journal of Rock Mechanics and Mining Sciences*, Vol. 190, p. 106100, Copyright (2025), published under a Creative Commons Attribution (CC BY 4.0) license https://creativecommons.org/licenses/by/4.0/)

and bond elastic modulus, cohesion, and tensile strength as main constitutive influencing elastic modulus and peak strength. SJM parameters such as tangential stiffness and joint tensile strength were highly influential on the shape of the post-peak response and localization of failure.

Yang et al. [46] investigated molecular dynamics simulation model that was developed using representative shale minerals (quartz, calcite, kaolinite, montmorillonite), as individual minerals and also as minerals mixed with kerogen. The tensile stress was applied in the models along the three crystallographic directions in order to estimate the directional mechanical behavior under temperature and pressure conditions

(360 K and 35 MPa, respectively) that replicated in-situ reservoir conditions in the Junggar Basin (Northwest China).

Tensile strength of mineral mixed with kerogen was observed to be lower than the counterparts with only mineral. There was an exception in case of kerogen-montmorillonite model which showed increased tensile strength and on the other hand kerogen-calcite model showed decrease in tensile strength. Their results emphasized the role of interfacial bonding which depending upon the mineral type and orientation either enhances or weakens the mechanical integrity. That study also assessed the impact of pore size and shape on tensile strength. They considered a variety of pores shapes (slit, square, cylindrical, and triangular pores) and sizes (2–10 nm) and developed models revealing that slit pores and triangular pores exhibit the highest resistance to tensile forces. Notably, larger pores resulted in a marked decrease in YM.

In order to produce true real-time simulations of reservoir behavior under a wide range of operational conditions, geomechanical models are required that can span the scale and resolution between seismic data, laboratory derived static mechanical properties, geologic interpretation and engineering design requirements. Xiao et al. [41] calculated the dynamic elastic properties and constructed 1D geomechanical models. These 1D models were then calibrated with the laboratory static mechanical properties derived from core samples and upscaled into 3D models using geostatistical techniques. Field-wide stress distribution were simulated using a finite element modelling (FEM) technique which incorporated both geomechanical and geological complexities (e.g., interbedded layers, faults and variable pore pressures). FEM was used in the calculation of detailed 3D stress tensors. The outputs of this model were then coupled with hydraulic-fracture and reservoir simulations to guide strategies for reservoir stimulation and design. The insights gained from such models provide a valuable approach for enhancing the link between subsurface geomechanical-geological characterization and engineering design.

5.4 Micro-Mechanical Characterization of Shales

Static tests are useful however conventional tests for static mechanical properties analysis requires large intact samples which is often difficult to obtain [27]. As a result, micromechanical-property characterization of shales via advanced techniques like nanoindentation and atomic force microscopy (AFM) has become an appealing option to help comprehend the mechanical heterogeneity of most shale formations [19].

Xie et al. [42] used nanoindentation in combination with scanning electron microscope (SEM) analysis to study the micromechanical behavior of laminated shales from the Ordos Basin (China) based on oriented structural characteristic. By combining SEM images and structural orientation entropy (SOE) the study quantitively assessed the grain orientation based on the size, shape, and spatial arrangement of mineral grains and pores. Highest elastic modulus was of pyrite whereas clay and

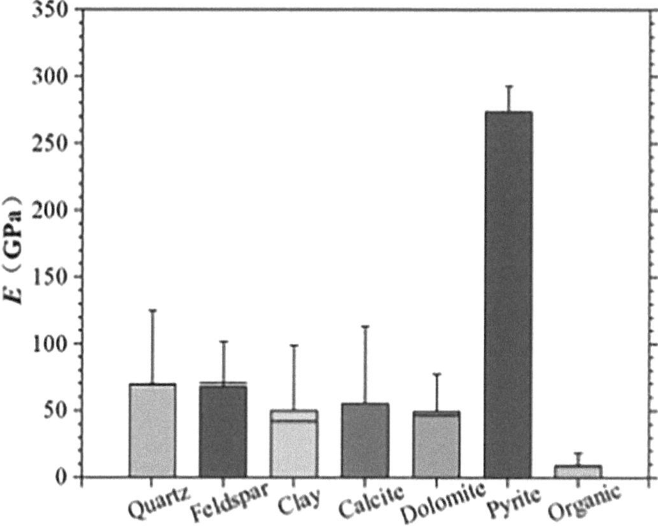

Fig. 5.8 Average elastic modulus of mineral grains in shale samples from Yan-Chang Formation, Ordos Basin. (Reprinted from *International Journal of Rock Mechanics and Mining Sciences*, Vol. 173, Xie, X., Deng, H., Hu, L., Li, Y., Mao, J. and Liu, J., "Assessing the effect of oriented structure characteristics of laminated shale on its mechanical behaviour with the aid of nano-indentation and FE-SEM techniques," p. 105625, Copyright (2024), with permission from Elsevier. License Number: 6038840491203)

organic phases showed lower elastic modulus. The individual grains showed a distinct strong gradient (Fig. 5.8) in elastic modulus, such that elasticity decreased from the geometric outwards to the edges. Using the Mori–Tanaka model, they upscaled the observed micromechanical properties to the mesoscale.

Liang et al. [17] used nano-indentation and AFM to study geomechanical characteristics evolution of shale samples during pyrolysis. The micromechanical study was integrated with Raman spectroscopy, Fourier transform infrared (FTIR) spectroscopy, and gas adsorption analyses to assess the impact of mineral composition, organic matter content, and pore evolution on the strength behavior of shale. Thermal maturation caused increase in Young's modulus of shale. S-shaped trend was observed for elastic modulus of organic matter with different peaks at mature and over-mature stages. Chemical changes in the kerogen caused these changes in the mechanical behavior. Loss of aliphatic chain resulted in dense packing of molecules during thermal maturation stage, whereas creation of graphite like structures due to aromatization at over-mature stage increased the stiffness of the sample.

The amount of quartz, increasing micropore surface area, and aromaticity with rising temperature correlated positively with an increase in elastic modulus. That study also developed a model explaining that the stiffness of shale initially increased due to compaction and early hydrocarbon generation. The shale samples then became stable due to pore expansion during peak hydrocarbon generation, before rising again in the over-mature stage of thermal maturity due to the development of graphite-like OM structure.

Alvayed et al. [1] observed dual-peak distribution of Young's modulus and hardness in shales. Organic matter and clays was the first phase, and the other phase represented harder minerals such as pyrite and quartz. The recorded nanoscale Young's modulus ranged between 7 to 85 GPa and hardness between 0.3 to 4.9 GPa. On the other hand, the macroscale Young's modulus ranged between 25 to 60 GPa. However, the mean values from both scales were closely aligned: nanoscale and macroscale moduli averaging 44 GPa and 42.6 GPa, respectively. These results demonstrate that the bulk mechanical behavior of shale can still be inferred from the micromechanical behavior given that the nanoscale measurements are sufficiently numerous to provide a statistically valid distribution of values.

In of underground CO_2 sequestration's context, the evaluation of CO_2-water interactions is important for long-term behavior of the reservoir. Liu et al. [18] assessed the micromechanical behavior of shales after CO_2-water-rock interaction of Marcellus Shale samples from Pennsylvania (U.S.A). They employed high-resolution nanoindentation and a TESCAN integrated mineral analyzer (TIMA) for determining mechanical properties of minerals present in the shale samples. They observed that CM exhibit a considerable reduction in mechanical stiffness (~15%) after CO_2-water saturation. On the other hand, no change in mechanical properties was observed for quartz. That study pioneered the application of accurate grain-based modelling (AGBM) to upscale the microscale properties, thereby facilitating the prediction of the macroscale mechanical properties. The actual mineral maps produced using TIMA were used to develop FEM models which enabled retention of the real shapes, sizes and distributions of mineral grains. The mechanical parameters derived from the nano-indentation tests were assigned to each mineral phase and simulation of uniaxial compression was performed to predict the bulk strength and elastic moduli.

5.5 Upscaling of Micromechanical Properties to Macro-Mechanical Properties

The problem with the micromechanical approach for determining the geomechanical properties of shales is the difficulty in estimating the macro-scale bulk mechanical properties from the micro-scale measurements. To bridge this gap several studies have applied upscaling methods to estimate bulk mechanical properties from nano-scale mechanical properties, as discussed in Sect. 5.4.

Sadeghpur et al. [26] through five upscaling models (Differential Effective Medium (DEM), Mori–Tanaka (MT), Self-Consistent Approximation (SCA), Kuster–Toksöz Formulation (KTF), and the Dilute Model (DM)) predicted YM of shales. It was observed that OM had a negative impact on YM and among the five upscaling methods, the DEM and DM methods generated results that were similar to the measured bulk mechanical values (Fig. 5.9).

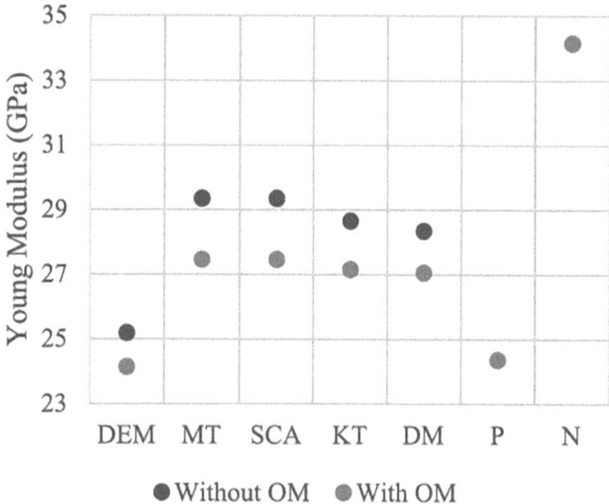

Fig. 5.9 Comparison of YM estimated from upscaling methods with and without OM. (Reprinted from Geoenergy Science and Engineering, Vol. 239, Sadeghpour, F., Darkhal, A., Gao, Y., Motra, H.B., Aghli, G. and Ostadhassan, M., "Comparison of geomechanical upscaling methods for prediction of elastic modulus of heterogeneous media," p. 212915, Copyright (2024), with permission from Elsevier. License Number: 6030210783347)

Wang et al. [33] correlated experimental observations with several theoretical upscaling models which yielded predictions on a large scale for both the nanoindentation-based phase data used and mineralogical compositions obtained from XRD and SEM–EDS analyses. The goal was to identify which upscaling method best replicates the mechanical response seen in large-scale triaxial tests. Unambiguous conclusion from these results was that YM values from nanoindentation were consistently much higher than those from triaxial compression tests and that the difference resulted from extremely high stiffness of isolated mineral phases and failure of nanoindentation to capture the expected contribution of pores and boundaries of grains. Out of the different upscaling approaches, the mineralogy-based MT and SC models had the highest Young's moduli (from ~36 to 58 GPa), and the nanoindentation-based MT model produced intermediate values (30–49 GPa). Of all models tested, the VRH produced the lowest estimates (21–35 GPa), but these numbers were closest to results from microindentation and triaxial tests. This indicates that although no technique is able to completely reproduce the macroscopic behavior, the VRH model and microindentation come the closest to realistic representations given current methodological limitations. According to that study, marlstone samples had consistently higher YM values than shale because of their calcite-rich matrix and lower clay content. These triaxial tests under several confining pressures achieved that YM increases as a linear function of the confining pressure. The increasing rate continuously decreases when the confining pressure grows large. This pressure dependence was stronger in shale than marlstone, showing the mineralogical composition controls the mechanical response to stress the most.

5.6 Predicting Geomechanical Properties in Wellbore Sections with Machine Learning

The capacity to predict geomechanical characteristics over a whole wellbore, or section of a reservoir, is essential to sustaining a high degree of safety during drilling [8], wellbore stability, and safe mud-weight windows [39], prediction of pore pressure [22], and for quantified evaluations of shale reservoirs on a large scale. However, whole-rock or side-wall core samples are never available to provide sufficient coverage to achieve this from laboratory measurements alone. Hence, it is necessary to estimate these properties using variables recorded while drilling (e.g., from mud-logs), and/or variables recorded by wireline or measurement-while-drilling petrophysical logging, either during or immediately following drilling. This is typically achieved using machine and/or deep learning techniques. Such methods require calibrations between dynamic values (recorded by downhole logs) and static values (measured under laboratory conditions) obtained from the few available cores from a formation, typically from surrounding wells [5].

Davoodi et al. [6] developed a model which predicted UCS using drilling/mud-log from two wells drilled in the Rag-e-Safid oil field (Iran) as input data. That model provided predictions for 500+ m drilled section. The predictions were calibrated from available laboratory measurements of core samples taken in surrounding wells. Davoodi et al. [7] demonstrated that a Gaussian process regression model could also accurately and continuously predict YM in real-time from the same two-well dataset calibrated with limited core data from surrounding wells.

A key application for geomechanical parameters in shale formation evaluations is the estimation of brittleness indices (BI) in 1-D (individual wellbores) and 3-D (field-wide distributions by integrating data from multiple wellbores and seismic datasets. There are two distinct shale BI that are commonly evaluated: one determined from mineralogy (*BIm*) defined in different ways using different shale mineral components [14, 31]; the other determined using geomechanical (elastic) rock properties (*BIg*) [10, 23]. Whereas *BIm* can be calculated using empirical formula calibrated with core data or information from specialist well logs capable of distinguishing mineralogy continuously through a wellbore (e.g., spectral gamma ray tool) [40], *BIg* calculations are typically derived using empirical formulas calibration with core measurements of elastic properties (Fig. 5.10). Common practice is to predict either or both BI indices with machine/deep learning methods using the available recorded petrophysical well-log data in multiple wellbores to provide 1D and 3D (field-wide) BI interpretations.

Verma et al. [30] predicted *BIm* using core and well-log data from two Barnett Shale (U.S.A) wells using artificial neural network (ANN) and support vector machine models. Wood [36] applied the transparent open box algorithm to the same two-well dataset demonstrating that *BIm* could be predicted more accurately and transparently with that machine-learning algorithm. Ye et al. [47] derived *BIg* from stress–strain tests applied to multiple cores from three Sichuan Basin (China) shales and successfully predicted those values from a standard suite of five well-logs

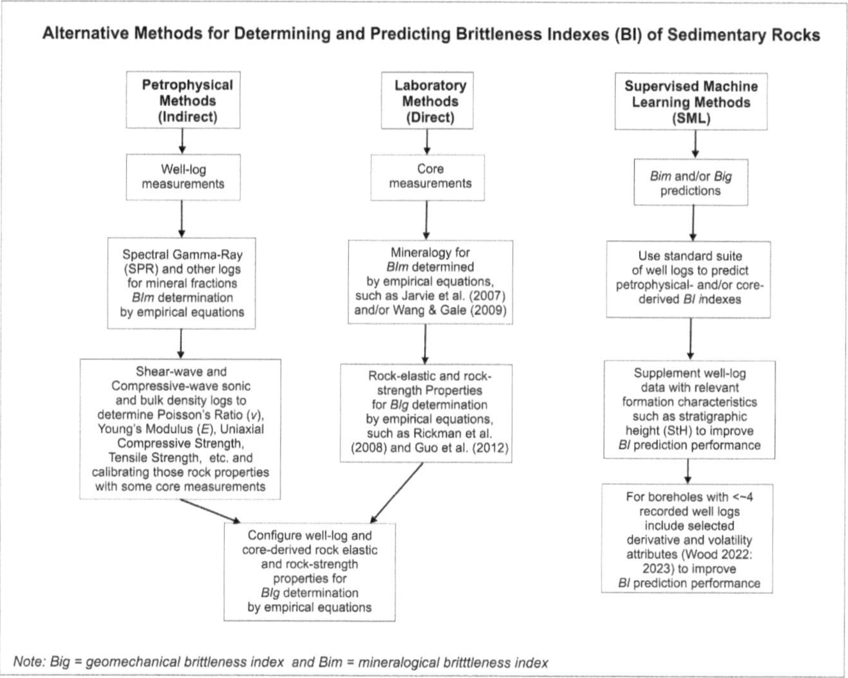

Fig. 5.10 Summary of information required to determine mineralogical (*BIm*) and geomechanical (*BIg*) brittleness indices in shales and other tight rock formations. (Reprinted from Implementation and Interpretation of Machine and Deep Learning to Applied Subsurface Geological Problems, David A. Wood, 'Predicting brittleness indexes in tight formation sequences,' pp. 79–109, Copyright (2025), with permission from Elsevier. License Number: 6030200198220)

using principal component analysis coupled with a back-propagating neural network model.

Ore and Gao [21] demonstrated that a gradient boosting model could successfully predict *BIg* derived from empirical geomechanical formulas in several Marcellus shale (U.S.A) wells using a suite of seven standard well-logs. Wood [38], using data from two of the wells from that study, demonstrated that an elastic net (multi-linear regression) model and an extreme gradient boosting model could separately generate reliable *BIg* predictions from just two or three recorded well-logs by calculating well-log attributes [37] from the recorded bulk density, compressional acoustic and gamma ray well logs.

The ability to predict *BIm*, *Big*, and potentially other geomechanical properties from just a few recorded well logs, calibrated with a few cores recovered from some wells, represents a substantial step forward. The development of shale reservoirs to produce oil and/or gas requires the drilling of many hundreds of closely spaced wellbores. It is not cost-effective to run a comprehensive suite of petrophysical well logs or recover cores from many of those wells. Most field development wells are drilled rapidly and record the most basic suite of well logs required for geo-steering and wellbore stability analysis (e.g., caliper, gamma ray, compressional acoustic and bulk density logs). Hence, to develop reliable 3D geomechanical property distributions across an entire shale reservoir it is important to be able to predict those properties from such limited suites of recorded well logs assisted by well-log attributes derived from those recorded logs.

References

1. Alvayed D, Alafnan S, Aljawad MS, Amao AO (2025) Integrative analysis of mechanical heterogeneity in organic-rich shales: bridging nano-and macroscale properties for an enhanced geomechanical understanding. Arab J Sci Eng 1–15
2. Amann F, Button EA, Evans KF, Gischig VS, Blümel M (2011) Experimental study of the brittle behavior of clay shale in rapid unconfined compression. Rock Mech Rock Eng 44:415–430
3. Chang C, Zoback MD, Khaksar A (2006) Empirical relations between rock strength and physical properties in sedimentary rocks. J Petrol Sci Eng 51(3–4):223–237
4. Chen D, Pan Z, Ye Z (2015) Dependence of gas shale fracture permeability on effective stress and reservoir pressure: model match and insights. Fuel 139:383–392. https://doi.org/10.1016/j.fuel.2014.09.018
5. Davarpanah SM, Van P, Vasorhelyi B (2020) Investigation of the relationship between dynamic and static deformation moduli of rocks. Geomech Geophys Geo-Energy Geo-Resour 6(1). https://doi.org/10.11007/s40948-020-00155-z. (Springer Science and Business Media Deutschland GmbH, Hungary)
6. Davoodi S, Mehrad M, Wood DA, Rukavishnikov VS, Bajolvand M (2023) Predicting uniaxial compressive strength from drilling variables aided by hybrid machine learning. Int J Rock Mech Min Sci 170:105546. https://doi.org/10.1016/j.ijrmms.2023.105546
7. Davoodi S, Mehrad M, Wood DA, Al-Shargabi MA, Eremyan G, Shulgina TA (2024) Novel data-driven model for real-time prediction of static Young's modulus applying mud-logging data. Earth Sci Inf 17:5771–5793. https://doi.org/10.1007/s12145-024-01474-5
8. Edan BK, Hussein HA (2023) Geomechanics analysis of well drilling instability: a review. J Eng 29(08):94–105. https://doi.org/10.31026/j.eng.2023.08.07
9. Feng X, Gong B, Liang Z, Wang S, Tang CA, Li H, Ma T (2024) Study of the dynamic failure characteristics of anisotropic shales under impact Brazilian splitting. Rock Mech Rock Eng 57(3):2213–2230
10. Guo ZQ, Chapman M, Li XY (2012) Correlation of brittleness index with fractures and microstructure in the Barnett shale. In: 74th European association of geoscientists and engineers conference and exhibition, incorporating SPE EUROPEC 2012: responsibly securing natural resources. European Association of Geoscientists and Engineers, EAGE, China
11. Guo T, Zhang S, Qu Z, Zhou T, Xiao Y, Gao J (2014) Experimental study of hydraulic fracturing for shale by stimulated reservoir volume. Fuel 128:373–380
12. Hazra B, Vishal V, Sethi C, Chandra D (2022) Impact of supercritical CO_2 on shale reservoirs and its implication for CO_2 sequestration. Energy Fuels 36(17):9882–9903
13. Heng S, Guo Y, Yang C, Daemen JJ, Li Z (2015) Experimental and theoretical study of the anisotropic properties of shale. Int J Rock Mech Min Sci 74:58–68
14. Jarvie DM, Hill RJ, Ruble TE, Pollastro RM (2007) Unconventional shale-gas systems: The Mississippian Barnett Shale of north-central Texas as one model for thermogenic shale-gas assessment, American Association of Petroleum Geologists Bulletin. US Am Assoc Pet Geol 91(4):475–499. https://doi.org/10.1306/12190606068
15. Josh M, Esteban L, Delle Piane C, Sarout J, Dewhurst DN, Clennell MB (2012) Laboratory characterisation of shale properties. J Petrol Sci Eng 88:107–124

16. Kahraman SAİR (2001) Evaluation of simple methods for assessing the uniaxial compressive strength of rock. Int J Rock Mech Min Sci 38(7):981–994
17. Liang Z, Jiang Z, Tang X, Chen R, Arif M (2024) Mechanical properties of shale during pyrolysis: atomic force microscopy and nano-indentation study. Int J Rock Mech Min Sci 183:105929
18. Liu Y, Liu Q, Wu Z, Liu S, Kang Y, Tang X (2024) Cross-scale mechanical softening of Marcellus shale induced by CO_2-water-rock interactions using nanoindentation and accurate grain-based modeling. Undergr Space 19:26–46
19. Ma Z, Pathegama Gamage R, Zhang C (2020) Application of nanoindentation technology in rocks: a review. Geomech Geophys Geo-Energ Geo-Resour 6:60. https://doi.org/10.1007/s40 948-020-00178-6
20. Olson JE (2008) Multi-fracture propagation modeling: applications to hydraulic fracturing in shales and tight gas sands. In: ARMA US rock mechanics/geomechanics symposium, pp ARMA-08. ARMA
21. Ore T, Gao D (2023) Prediction of reservoir brittleness from geophysical logs using machine learning algorithms. Comput Geosci 171:105266. https://doi.org/10.1016/j.cageo.2022.105266
22. Radwan AE, Wood DA, Radwan AA (2022) Machine learning and data-driven prediction of pore pressure from geophysical logs: a case study for the Mangahewa gas field, New Zealand. J Rock Mech Geotech Eng 14(6):1799–1809. https://doi.org/10.1016/j.jrmge.2022.01.012
23. Rickman R, Mullen MJ, Petre JE, Grieser WV, Kundert D (2008) A practical use of shale petrophysics for stimulation design optimization: all shale plays are not clones of the Barnett Shale. In: Proceedings of the SPE annual technical conference and exhibition, society of petroleum engineers, SPE, p 115258. https://doi.org/10.2118/115258-MS
24. Rybacki E, Meier T, Dresen G (2016) What controls the mechanical properties of shale rocks?–Part II: Brittleness. J Petrol Sci Eng 144:39–58
25. Rybacki E, Reinicke A, Meier T, Makasi M, Dresen G (2015) What controls the mechanical properties of shale rocks?–Part I: strength and Young's modulus. J Petrol Sci Eng 135:702–722
26. Sadeghpour F, Darkhal A, Gao Y, Motra HB, Aghli G, Ostadhassan M (2024) Comparison of geomechanical upscaling methods for prediction of elastic modulus of heterogeneous media. Geoenergy Sci Eng 239:212915
27. Sethi C, Hazra B, Wood DA, Singh AK (2023) Experimental protocols to determine reliable organic geochemistry and geomechanical screening criteria for shales. J Earth Syst Sci 132(2):45
28. Sethi C, Motra HB, Hazra B, Ostadhassan M (2025) Influence of lithological contrast on elastic anisotropy of shales under true-triaxial stress and thermal conditions. Int J Rock Mech Min Sci 190:106100
29. Sone H, Zoback MD (2013) Mechanical properties of shale-gas reservoir rocks—part 1: static and dynamic elastic properties and anisotropy. Geophysics 78(5):D381–D392
30. Verma S, Zhao T, Marfurt KJ, Devegowda D (2016) Estimation of total organic carbon and brittleness volume. Interpretation 4(3):T373–T385
31. Wang FP, Gale JFW (2009) Screening criteria for shale-gas systems. Gulf Coast Assoc Geol Soc Trans 59:779–793
32. Wang J, Yang C, Guo Y, Liu Y, Jiang W, Luo Y, Wu Y, Xiong Y, Peng PA (2025) Examining the reliability of current micro-and nano-indentation-based rock mechanical upscaling schemes: a comprehensive comparison with uniaxial/triaxial macroscopic mechanical testing. Geomech Geophys Geo-Energy Geo-Resour 11(1):28
33. Wang L, Wang G, Yang D, Zhao J, Kang Z, Zeng Q, Zhao Y (2025) Research on the mechanical properties and anisotropy evolution of uniaxial compression of oil shale under real-time high-temperature steam. Rock Mech Rock Eng 58(2):1911–1932
34. Wang W, Wang X, Li B, Yang Z (2024) Study on parallel bond and smooth joint dual model parameter calibration of mechanical properties of shale under pressure. Rock Mech Rock Eng 57(11):9445–9475
35. Wood DA, Hazra B (2017) Characterization of organic-rich shales for petroleum exploration & exploitation: a review-part 3: applied geomechanics, petrophysics and reservoir modeling. J Earth Sci 28:779–803. https://doi.org/10.1007/s12583-017-0734-8

36. Wood DA (2021) Brittleness index predictions from Lower Barnett Shale well-log data applying an optimized data matching algorithm at various sampling densities. Geosci Front 12(6):101087. https://doi.org/10.1016/j.gsf.2020.09.016. (Elsevier BV)
37. Wood DA (2022) Predicting brittleness indices of prospective shale formations from sparse well-log suites assisted by derivative and volatility attributes. Adv Geo-Energy Res 6(4):334–346. https://doi.org/10.46690/ager.2022.04.08
38. Wood DA (2023) Geomechanical brittleness index prediction for the Marcellus Shale exploiting well-log attributes. Results Eng 100846. https://doi.org/10.1016/j.rineng.2022.100846
39. Wood DA (2024) Wellbore stability and the establishment of a safe mud weight window. In: Wood DA, Cai J (eds) Sustainable natural gas drilling: technologies and case studies for the energy transition, pp 135–168. Elsevier. https://doi.org/10.1016/B978-0-443-13422-7.000 09-X. (Chapter 5)
40. Wood DA (2025) Predicting brittleness indexes in tight formation sequences. In: Implementation and interpretation of machine and deep learning to applied subsurface: problems prediction models exploiting well-log information, pp 79–109. Elsevier. https://doi.org/10.1016/B978-0-443-26510-5.00003-7. (Chapter 3)
41. Xiao Y, Liang C, Zhu D, Zou C, Yan J, Bai Y (2024) Multi-scale geomechanical modelling of unconventional shale gas: the implication on assisting geophysics–geology–engineering integration. Int J Energy Res 2024(1):4145930
42. Xie X, Deng H, Hu L, Li Y, Mao J, Liu J (2024) Assessing the effect of oriented structure characteristics of laminated shale on its mechanical behaviour with the aid of nano-indentation and FE-SEM techniques. Int J Rock Mech Min Sci 173:105625
43. Xu R, Gao L, Jin Y, Wang Y, Li Z, Hu S (2025) Influence of strain measurement methods on crack initiation and crack damage thresholds of shale. Geomech Geophys Geo-Energy Geo-Resour 11(1):34
44. Xue L, Qin S, Sun Q, Wang Y, Lee LM, Li W (2014) A study on crack damage stress thresholds of different rock types based on uniaxial compression tests. Rock Mech Rock Eng 47:1183–1195
45. Yan D, Zhao L, Wang Y, Zhang Y, Cai Z, Song X, Zhang F, Geng J (2023) Heterogeneity indexes of unconventional reservoir shales: quantitatively characterizing mechanical properties and failure behaviors. Int J Rock Mech Min Sci 171:105577
46. Yang X, Chen J, Zhang X, Jiang F, Pang H (2025) Influence of the mineral composition and pore structure on the tensile mechanical properties of shale from a microscopic perspective: insights from molecular dynamics simulations. Int J Rock Mech Min Sci 191:106123
47. Ye Y, Tang S, Xi Z, Jiang D, Duan, Y., 2022. A new method to predict brittleness index for shale gas reservoirs: Insights from well logging data. J Pet Sci Eng 208:109431. https://doi.org/10.1016/j.petrol.2021.109431. (Elsevier BV)
48. Zhang R-H, Zhang L-H, Wang R-H, Zhao Y-L, Zhang D-L (2016) Research on transient flow theory of a multiple fractured horizontal well in a composite shale gas reservoir based on the finite-element method. J Nat Gas Sci Eng 33:587–598. https://doi.org/10.1016/j.jngse.2016.05.065

Chapter 6
Research Gaps and Outstanding Challenges

Abstract Accurate characterization of shale formations still faces some key challenges due to their complex matrix heterogeneity and fabric anisotropy. Predicting the mechanical behavior of shales using geochemical proxies has generated promising provisional results. However, challenges including compositional variation and data resolution need to be overcome. Imaging techniques such as correlative microscopy and X-ray CT are effective at characterizing multi-scale shale properties. However, the integration of both data and sample preparation techniques is cumbersome and requires improvement and simplification. Pore-structure characterization determined using mercury intrusion and nuclear magnetic resonance techniques needs to be better calibrated to resolve disparities due to fundamental differences in the measurement techniques. Geomechanical property prediction is currently constrained by a proper micromechanical upscaling models. For better representation of actual shale microstructure, further advancements in digital rock modelling and image-based simulations are needed. To further improve the evaluation and development of shale reservoirs, future research should be aimed at bridging these gaps through the advancement of more integrated techniques and interdisciplinary approaches.

Keywords Well-log interpretation · Machine learning · Geomechanical upscaling · Correlative microscopy · X-ray computed tomography (CT) · Digital rock modelling

6.1 Geochemical Characterization

Recent developments that integrate X-ray diffraction (XRD) and X-ray fluorescence (XRF) data [5, 6] can substantially improve the mineralogical characterization of shales. However, some challenges remain warranting further research, especially when dealing with variable mineral stoichiometries or subtle compositional gradients within mineral grains and organic matter particles. The elemental distributions

C. Sethi et al., *Recent Advances in Shale Characterization Based on Laboratory Geochemistry and Geomechanics Techniques*, SpringerBriefs in Petroleum Geoscience & Engineering, https://doi.org/10.1007/978-3-032-03961-3_6

detected by XRF can be affected by these variations which may lead to uncertainties when correlating specific elemental concentrations to particular mineral phases. Furthermore, as shales are highly anisotropic and show spatial heterogeneity in three dimensions, the full extent of compositional variations may not be fully detected if sample resolution is low, or the number of samples assessed is too few.

Prediction of geomechanical properties using geochemical proxies is a realistic trajectory as evidenced by a growing body of literature [7, 10]. Mineral composition, depositional conditions, and organic content all have direct effects on rock strength and brittleness. These influences can be quite different from one formation to another based on lithological and paleoenvironmental considerations. Development of universally applicable geochemical proxy models held great potential but is a pursuit deserving of renewed research effort. In order to push predictive modelling in machine-learning models, it's vital for these models to be trained on large, diverse datasets, which will lead to more reliable predictions on mechanical behavior of shale, using geochemical proxies as inputs. The machine-learning techniques recently developed and applied to predict geomechanical properties from well logs [2] provide exciting and powerful methods that should be adapted and applied to large scale geochemical datasets.

6.2 Multi Scale Imaging

Imaging techniques such as optical-electron correlative microscope (OECM) that integrates different spatial resolutions of optical and electron microscopes for petrographic and microstructural characterization of shale organic matter (OM) have advanced in recent years [4, 11, 14]. However, the OM marked as region of interest (ROI) under optical microscope and its subsequent relocation under scanning electron microscopy (SEM) is a technically demanding and time-consuming process. Factors like sample polishing quality, removal of immersion oil after optical microscopy, gold coating of samples before SEM imaging can adversely affect ROI relocation. SEM parameters (e.g., acceleration voltage and working distance) affect contrast and help in the discernibility of organic phase and mineral phase in shales. Yet further work is needed to determine best practice conditions through a range of shale compositions and thermal maturity stages. Also, great potential is there if constructing digital cores from FIB-SEM datasets is considered [20] for shales using more advanced convolutional neural network algorithms to further enhance the classification of inorganic minerals and organic fragments within shales.

X-ray computed tomography (micro-CT and nano-CT) measurements of pore morphology and connectivity are scale-dependent [9, 12, 21]. Between micro-CT and nano-CT, the measurement results of pore throat diameter, aspect ratio and tortuosity may vary significantly. Such variations can complicate the integration of multi-scale porosity datasets into digital-rock-physics models. Further work is required to improve the integration of multi-scale datasets in ways that mitigate scale-dependency issues.

6.3 Multi Scale Characterization of Shale Petrophysical Properties

Recent advances using mercury injection capillary pressure (MICP), low-pressure gas adsorption (LPGA) and CO_2-based fractal analysis have expanded the capabilities of characterizing shale formations based on their petrophysical properties. Essential corrections need to be applied to MICP data to achieve conformance with respect to volume and grain compressibility [1]. However, the detailed definition and application of such corrections relating to certain lithologies is yet to be completed and verified, requiring future research to do so. Further scope also exists to expand the use of lacunarity and succolarity measurements derived from LPGA analysis [13] to expand the heterogeneity and anisotropy characterization of shale pore-sized distributions in three dimensions.

MICP integration with nuclear magnetic resonance (NMR) provides important insights into pore-body and pore-throat connectivity [3]. The measurement principle of MICP relies on pressure-driven fluid invasion. On the other hand, NMR captures magnetic relaxation in fluid saturation conditions to determine petrophysical properties from scanned or logged samples. Thus, MICP inherently only measures the pores which are accessible by mercury invasion, and its results are affected by the non-wetting nature of mercury, geometry and surface roughness of the pores, and mechanical deformation under high intrusion pressure. In contrast, NMR is a non-destructive method, which is highly sensitive to surface-to-volume ratios, making it suitable for capturing microporosity details. The fundamentally different physical and measurement conditions of these two methods can be challenging when attempting to integrate information generated by each of them for the same sample. To fully benefit from the complementary nature of data generated using these two methods, more reliable calibration strategies are required. Future research should be focused on improving the alignment of MICP- and NMR-derived pore-size interpretations exploiting artificial pore-network modelling, and/or imaging-derived pore geometries. There is further scope to improve upon the recent convolutional neural network models applied to field emission scanning electron microscopy (FIB-SEM) datasets, such as KiU-net [19] to assess micro-pore networks.

6.4 Geomechanical Characterization of Shales

Significant advances in the geomechanical characterization of shales through both experimental and modelling approaches have been observed in the recent years. As these techniques evolve, some challenges still exist that provides prospects for additional study. The measurement of axial strains using different techniques, such as extensometer and linear variable differential transformer (LVDT), influence the stress threshold values determined for anisotropic shales [18]. These two methods are both widely accepted but generate distinctive post-peak strain behavior and elastic behavior. To reconcile these differences, it is necessary to consider the resolution of the strain measurements determined and the spatial coverage involved. Hence, technique enhancements are required to incorporate resolution in recorded extensometer and LVDT measurements to improve the characterization of strain responses of the rock samples tested.

Real-time thermal treatment during uniaxial loading of shale samples has revealed a distinctive transitional trend as temperature rises changing from brittle to ductile behavior but reverting back to brittle behavior as temperatures rise above 400 °C [16]. This finding has particular relevance for evaluating the suitability of subsurface shale formations for exploitation as geothermal or CO_2-sequestration scenarios. Further research is required to quantify shale brittleness/toughness with temperature in relation to other influencing factors such as real-time injection of CO_2 involving cyclic or prolonged heating episodes.

There is substantial scope to expand the use of petrophysical well logs, calibrated with limited core data, to provide continuous, predicted wellbore profiles of many geomechanical shale properties of interest. Combining such profiles for multiple wellbore across a shale formation has the potential to provide 3-D models with comprehensive spatial coverage, especially when integrated with seismic data. However, a key limitation to this approach is that many wellbore record only a few key petrophysical logs to save cost and time. The recently developed technique of calculating certain well-log attributes to predict shale mineralogical and geomechanical brittleness indices [17] could be further exploited to predict other geomechanical properties of interest from sparse suites of available well logs.

Upscaling of micromechanical parameters to predict the bulk mechanical properties of shales has been recently achieved using theoretical homogenization models including the differential effective medium (DEM), Mori–Tanaka (MT), self-consistent approximation (SCA), and Voigt-Reuss-Hill (VRH) models (Sadeghpur et al. 2024) [8, 15]. These models have been applied to derive stiffness values from nano-indentation and mineralogical composition data determined by XRD and SEM–EDS analyses. These homogenization models rely on simplification of the assumptions for grain behavior and grain interactions by treating mineral grains as only spherical and ellipsoidal inclusions distributed in the shale matrix, and assuming perfect bonding between phases. However, in most shale formations the matrix is composed of irregular grain shapes, imperfect grain contacts with some micropores present at grain boundaries (intergranular porosity). Based on such assumptions, DEM, MT, and VRH models can only provide first-order approximations of the geomechanical properties they evaluate.

For the above-mentioned reasons, the predictive capacity of theoretical homogenization models is uncertain for highly heterogenous, anisotropic rocks, such as those comprising most shale formations. Future research is required to seek alternative micro-mechanical-property upscaling methods that should include image-based modelling based on 3D digital reconstructions depicting actual shale microstructures derived from high-resolution X-ray CT, focused ion beam SEM (FIB-SEM), such as those used to generate digital cores. Such improved techniques should allow for more accurate representation of grain geometry, phase distribution and pore connectivity which should generate more reliable bulk mechanical behavior from micromechanical data. To verify their reliability, it would be desirable to calibrate their results against experimental data for multiple samples derived from true-triaxial compression tests replicating in-situ temperature and pressure conditions.

References

1. Davudov D, Moghanloo RG, Lan Y (2018) Evaluation of accessible porosity using mercury injection capillary pressure data in shale samples. Energy Fuels 32(4):4682–4694
2. Davoodi S, Mehrad M, Wood DA, Rukavishnikov VS, Bajolvand M (2023) Predicting uniaxial compressive strength from drilling variables aided by hybrid machine learning. Int J Rock Mech Min Sci 170:105546. https://doi.org/10.1016/j.ijrmms.2023.105546
3. Gu Z, Wang S, Guo P, Zhao W (2024) Utilizing differences in mercury injection capillary pressure and nuclear magnetic resonance pore size distributions for enhanced rock quality evaluation: a Winland-style approach with physical meaning. Appl Sci 14(5):1881
4. Hazra B, Singh PK, Sethi C, Pandey JK (2024) Application of optical-electron correlative microscopy for characterization of organic matter. J Geol Soc India 100(10):1385–1394
5. Hupp BN, Donovan JJ (2018) Quantitative mineralogy for facies definition in the Marcellus Shale (Appalachian Basin, USA) using XRD-XRF integration. Sed Geol 371:16–31
6. Kim JJ, Ling FT, Plattenberger DA, Clarens AF, Lanzirotti A, Newville M, Peters CA (2021) SMART mineral mapping: synchrotron-based machine learning approach for 2D characterization with coupled micro XRF-XRD. Comput Geosci 156:104898
7. Li H, Li D, He Q, Sun Q, Zhao X (2024) Controlling mechanism of shale palaeoenvironment on its tensile strength: a case study of Banjiuguan Formation in Micangshan Mountain. Fuel 355:129505
8. Sadeghpour F, Darkhal A, Gao Y, Motra HB, Aghli G, Ostadhassan M (2024) Comparison of geomechanical upscaling methods for prediction of elastic modulus of heterogeneous media. Geoenergy Sci Eng 239:212915
9. Saif T, Lin Q, Gao Y, Al-Khulaifi Y, Marone F, Hollis D, Blunt MJ, Bijeljic B (2019) 4D in situ synchrotron X-ray tomographic microscopy and laser-based heating study of oil shale pyrolysis. Appl Energy 235:1468–1475
10. Sethi C, Hazra B, Ostadhassan M, Motra HB, Dutta A, Pandey JK, Kumar S (2024) Depositional environmental controls on mechanical stratigraphy of Barakar Shales in Rajmahal basin, India. Int J Coal Geol 285:104477
11. Sethi C, Mastalerz M, Hower JC, Hazra B, Singh AK, Vishal V (2023) Using optical-electron correlative microscopy for shales of contrasting thermal maturity. Int J Coal Geol 274:104273
12. Sun M, Fu J, Wang Q, Gao Z (2024) Pore network characterization and fluid occurrence of shale reservoirs: state-of-the-art and future perspectives. Adv Geo-Energy Res 12(3):161–167
13. Tian Z, Wei W, Zhou S, Wood DA, Cai J (2021) Experimental and fractal characterization of the microstructure of shales from Sichuan basin, China. Energy & Fuels 35(5):3899–3914. https://doi.org/10.1021/acs.energyfuels.0c04027
14. Valentine BJ, Hackley PC (2019) Applications of correlative light and electron microscopy (CLEM) to organic matter in the North American shale petroleum systems
15. Wang J, Yang C, Guo Y, Liu Y, Jiang W, Luo Y, Wu Y, Xiong Y, Peng PA (2025) Examining the reliability of current micro-and nano-indentation-based rock mechanical upscaling schemes: a comprehensive comparison with uniaxial/triaxial macroscopic mechanical testing. Geomech Geophys Geo-Energy Geo-Resour 11(1):28
16. Wang L, Wang G, Yang D, Zhao J, Kang Z, Zeng Q, Zhao Y (2025) Research on the mechanical properties and anisotropy evolution of uniaxial compression of oil shale under real-time high-temperature steam. Rock Mech Rock Eng 58(2):1911–1932

17. Wood DA (2025) Predicting brittleness indexes in tight formation sequences. In: Implementation and interpretation of machine and deep learning to applied subsurface: problems prediction models exploiting well-log information, pp 79–109. Elsevier. https://doi.org/10.1016/B978-0-443-26510-5.00003-7. (Chapter 3)
18. Xu R, Gao L, Jin Y, Wang Y, Li Z, Hu S (2025) Influence of strain measurement methods on crack initiation and crack damage thresholds of shale. Geomech Geophys Geo-Energy Geo-Resour 11(1):34
19. Yasin Q, Liu B, Sun M, Ismail A, Wood DA, Tian X, Yan B, Fu L, Fu X (2024) An improved convolutional architecture for pore structure analysis in organic-rich shale using FIB-SEM. Int J Coal Geol 295:104625. https://doi.org/10.1016/j.coal.2024.104625
20. Yasin Q, Liu B, Sun M, Sohail SM, Asmail A, Wood DA, Yang Z (2025) Digital core modeling for multimineral segmentation of lacustrine shale oil using FE-SEM and KiU-Net. Fuel 398:135474. https://doi.org/10.1016/j.fuel.2025.135474
21. Zhao H, Liang B, Sun W, Hu Z, Ma Y, Liu Q (2022) Effects of hydrostatic pressure on hydraulic fracturing properties of shale using X-ray computed tomography and acoustic emission. J Petrol Sci Eng 215:110725

MIX
Papier aus verantwortungsvollen Quellen
Paper from responsible sources
FSC® C105338

If you have any concerns about our products,
you can contact us on
ProductSafety@springernature.com

In case Publisher is established outside the EU,
the EU authorized representative is:
Springer Nature Customer Service Center GmbH
Europaplatz 3, 69115 Heidelberg, Germany

Printed by Libri Plureos GmbH
in Hamburg, Germany